100 CAR FIXES

Car Problems Quick Solutions

By

SIMFUKWE ALFRED

ISBN: 978-99968-71-60-3

Published by Simfukwe Alfred Plot 13933, Maipei Ward, Mogoditshane, Botswana
PO BOX 76682960, River Walk, Gaborone, Botswana
Email: simfukweebooks@gmail.com

Email: simfukweebooks@gmail.com

TABLE OF CONTENTS

DEDICATION

It was around 11 am while I was busy doing leg squats in the gym when my phone started ringing. Looked on the phone to see what name was reflecting on the incoming call. The appeared name was Isaac. "I have an issue here. I am working on the ford focus right now at the client's premise. The car is just cranking and not STARTING.I have two computer boxes (ECU) for this car but none of them is switching ON the car. When I connect the diagnostic equipment to scan the computer box, the scanner is not able to read the computer box but it can read all the other control modules on the car. This is happening to both of these computer boxes. I don't think they can all be dead computer boxes. I have downloaded the pin connector data of the terminal connector of the computer box (ECU) from the internet. According to the data, the computer box is supposed to have two 12-volt positives. One 12 volts is supposed to be in a small wire while the other 12 volts in the big wire. After turning ON the ignition switch, I am able to test 12 volts positive in the small wire but in the big wire there is no 12 volts." Isaac narrated to me.

Isaac, here is what you can do: Trace the big wire and see if it is coming from the fuse box. If the wire is traced and found to be coming from the fuse box then it will mean that, that wire is supposed to have the 12 volts positive.

This is because in most cases the computer box (ECU) is supplied with two 12 volts positives, one is constant 12 volts while the other is ignition switch ON 12 volts." Ok.I will call you again let me do that" Isaac spoke and hung up the phone. After sometime, my phone rang again. It's Isaac. "I traced the big wire and it led me to the fuse box. This wire is connected to one of the busbars on the fuse box"Ok,that big wire is then supposed to be the one supplying constant 12 volts to the computer box (ECU) but seems like it is broken somewhere along the wiring harness.

That is why you cannot get 12 volts on it, on the terminal connector when you test.Create a long jumper wire that will have a fuse in between and then connect this wire to that big wire somewhere near the computer box then the other end of the wire connect it on the battery positive terminal. Then connect everything back the way they were including the computer box (ECU) and START the car." Ok, let me do that. I will call you". After sometime passed, then Isaac called. "The car has just STARTED" Ah, ok.So what did you do. "I connected the jumper wire the same way you said I should". This book is dedicate to ISAAC.

PREFACE

The majority of the problems that occur on cars can be fixed quickly if approached in a logical order. Car problems can range from simple to very difficult depending on how they are understood. A simple problem can for example be: battery terminals corroded and causing the car not to crank.

One day somebody that I know called me on the phone and asked what he should do as his car was not cranking. He had just parked the car on the public car parking lot and walked into the shop. Upon coming from the shop, the car refused to crank. He asked me what could be the problem. I suggested to him that he should disconnect the battery terminals and wait for ten minutes then reconnect the battery and START the car. He did that and after ten minutes the car STARTED.I don't really know what happened why the car did not crank just after parking it. However I just guessed based on my experience that maybe his battery terminals were corroded. By removing the battery terminals he probably indirectly created some good contact between the battery posts and the terminals.

Some car problems may be similar or same from one car to the other. It's not a bad idea to approach one problem the same way you approached the other problem of the same or similar car. Sometimes fixing the car based on how the similar car with the similar symptoms was fixed can save you time. Though sometimes same symptoms might not mean same problem.

The 100 CAR FIXES displays real problems that occurred on different cars and the fixes which were done. It takes you through the quick procedure of how the problems on different cars were fixed. The fixes on all the vehicles mentioned were done by the author of this book. There are so many ways of fixing car problems, therefore the author is not administering on how you should fix the cars but instead is sharing with you on how he solved the car problems mentioned in this book. The procedures which were used to fix the cars are shared so that someone might adopt them.

CHAPTER 1

ENGINE PROBLEMS

(1) CAR STARTS BUT SWITCHES OFF AFTER A SHORT WHILE

- Vehicle Make: VW
- Model: Polo (9N1/9N2/9N3)
- Year: 2002-2009
- Engine Code: BLM
- Steering Wheel: Right-Hand Drive
- Transmission: Automatic

PROBLEM DESCRIPTION BY VEHICLE DRIVER

The car starts, but the idling is at very unstable rpms. As it idles it will then just go OFF .Sometimes it can start and drive but will switch OFF along the way.

SYMPTOMS

- Very low rpms
- Unstable rpms
- Trouble codes:
    - P1128, Bank 1 mixture adaptation, multi-system too lean
    - P1136, Bank 1 mixture adaptation, system too lean

HOW FAULT FINDING WAS DONE

- Disconnected the discharge pipe from the fuel pump
- Connected the fuel pressure gauge onto the fuel pump discharge outlet
- Asked someone to crank engine while observing the fuel pressure gauge. The pressure gauge measured 0.5 bars. The correct pressure is supposed to range between 2.5 bars to 3 bars, therefore the fuel pump motor was condemned to be faulty.

WHAT CAUSED THE FAULT

- Tear and wear of the fuel pump motor

THE FIX

- Replaced the fuel pump motor with the new one

OPTIONAL FIX

- Replace the complete fuel pump unit including the fuel pump housing

(2) NOT REVVING UP

- Vehicle Make: Toyota
- Model: Vitz
- Year:2006-2011
- Engine Code: 1KR
- Steering Wheel: Right Hand Drive
- Transmission: Automatic

PROBLEM DESCRIPTION BY VEHICLE DRIVER

The car is able to start but cannot rev up. When DRIVE is selected and the accelerator pedal is depressed so that the car moves, it will drive very slowly and will not accelerate.

SYMPTOMS

- The idling is ok at its normal idling rpm
- The car can start very well
- When the accelerator is depressed fully, the maximum rpm is below 3000

HOW FAULT FINDING WAS DONE

- Diagnosed the car with a diagnostic equipment.
  Trouble code stored: P2118, Throttle motor current range/performance

- Tested wires to the throttle body. Wires to throttle body proved to be okay, therefore the throttle body condemned to be faulty

WHAT CAUSED THE FAULT

Tear and wear of throttle body

THE FIX

- Replaced the throttle body

(3) CRANKS BUT NO START

- Vehicle Make: Mercedes-Benz
- Model: M-CLASS
- Year: 1998-2005
- Engine Code: 112.942
- Steering Wheel: Right Hand Drive
- Transmission: Automatic

PROBLEM DESCRIPTION BY VEHICLE DRIVER

Before the car was parked it was just fine. After parking the car overnight, suddenly it refused to start in the morning.

SYMPTOMS

- The vehicle is able to crank but does not start

HOW FAULT FINDING WAS DONE

- Disconnected the fuel pipe from the injector rail. Placed the disconnected pipe into a transparent bottle and asked someone to crank the car while I observed if fuel will discharge into the transparent bottle. No fuel was discharged upon cranking.

- Tested for power supply and ground to the fuel pump. Power supply and ground proved to be ok, therefore the fuel pump motor condemned to be faulty. The fuel pump is located under the left rear passenger seat.

WHAT CAUSED THE FAULT

- Tear and wear of the fuel pump motor

THE FIX

- Replaced the complete unit of the fuel pump with a new one

(4) CRANKS BUT NOT STARTING

- Vehicle Make: Nissan
- Model: Hardbody
- Year: 1998-2002
- Engine Code: KA24E
- Steering Wheel: Right Hand Drive
- Transmission: Manual

PROBLEM DESCRIPTION BY VEHICLE DRIVER

The vehicle can only crank but can't start

SYMPTOMS

- Cranks only
- Dashboard lights come ON as normal

HOW FAULT FINDING WAS DONE

- Tested for spark while cranking (ok)
  Tested for injector pulse while cranking (ok)
- Disconnected the fuel delivery pipe and tested for pressure with a fuel pressure gauge. Pressure was less than 1 bar,
- system pressure is supposed to be 2.94 bars. Hence the fuel pump condemned to be faulty.

WHAT CAUSED THE FAULT

- Tear and wear of the fuel pump

THE FIX

- Replaced the fuel pump motor

OPTIONAL FIX

- Replace the complete fuel pump unit

(5) CAR LOSES POWER WHILE DRIVING

- Vehicle Make: VW
- Model: Polo (6R1)
- Year: 2009-2014

- Engine Code: CGG
- Steering Wheel: Right Hand Drive
- Transmission: Automatic

PROBLEM DESCRIPTION BY VEHICLE DRIVER

The car used to be very fine but recently it started this problem where by it will just lose the power while driving. When it loses the power it will become limited to below 2500 RPM. Then the car will be moving very slow only in gear one. Sometimes when the problem happens I will switch the car OFF. When I switch it OFF and START it immediately, then the car will drive just fine but just for a few minutes then again it will lose the power and only move in gear one.

SYMPTOMS

- Car STARTS but will soon lose power while driving
- The EPC light and the electronic stability programme symbol appears on the dashboard and also the tyre flat symbol, then car immediately loses the power
- RPM limited to below 2500 RPM

HOW FAULT FINDING WAS DONE

- Scanned the ECM with diagnostic scanner and retrieved trouble code: P0685 – ECM/PCM power relay control circuit open

- Inspected fuse F35 (15A) on the fuse box, the fuse box under the steering wheel. The fuse F35 was found to be bridged with a loose copper wire on its legs. The previous mechanic bridged the fuse since it blew up. The jumper wire was loose on the fuse legs hence causing the ECM to detect high resistance in the wire supplying power to the ECM.Fuse F35 is the fuse supplying ignition ON power to the ECM

WHAT CAUSED THE FAULT

Fuse jumped with loose wire

THE FIX

- Replaced the fuse F35 with a good 15A fuse
- IMPORTANT: Never jump a fuse with wire, just replace it…

(6) ONLY CRANKS

- Vehicle Make: Toyota
- Model: Camry
- Year: 2001 - 2005
- Engine Code: 2AZ-FE
- Steering Wheel: right-hand drive
- Transmission: Automatic

PROBLEM DESCRIPTION BY VEHICLE DRIVER

While driving, the accelerator pedal stopped responding and slowly the rev count dropped till the car stopped moving.

SYMPTOMS

- Only cranks
- Dashboard lights come ON as normal

HOW FAULT FINDING WAS DONE

- Disconnected outlet pipe from the fuel pump Connected fuel
- pressure gauge on the fuel pump outlet to test pressure while someone cranked the engine. The pressure was less than 1 bar. System pressure is supposed to be minimum of 3 bars. Hence
- fuel pump condemned to be faulty.

WHAT CAUSED THE FAULT

Tear and wear of the fuel pump

THE FIX

Replaced the fuel pump motor

OPTIONAL FIX

- Replace the complete fuel pump unit

(7) MISFIRING AND ERRATIC IDLING

- Vehicle: Make Mazda
- Model: Mazda 3 (Axela)
- Year: 2003-2009
- Engine Code: Z6
- Steering Wheel: Right Hand Drive
- Transmission: Automatic

PROBLEM DESCRIPTION BY VEHICLE DRIVER

Car started misfiring and would not idle properly

SYMPTOMS

- Rough idling
- Check Engine light showing on dashboard

HOW FAULT FINDING WAS DONE

- Trouble coderetrieved:
    - P0352-ignition coil primary or secondary circuit.
- Testedforpower supply to all the 4 coils (ok)
- Testedthesignal wires on all the 4 coils while car is on idle (ok)
- Disconnectedconnector to ignition coil 1 while car is on idle. No change tothe way car is idling.
- Swapped ignition coil 1 with ignition coil 2 while car is OFF. Then disconnected the connector to ignition coil 1 while car is on idle. The sound of the idling changed as the connector was removed.
- Disconnected connector from ignition coil 2 while car is on idle, no change to the sound of idling. Hence the coil which was initially on 1 and now on 2 is condemned to be faulty.

WHAT CAUSED THE FAULT

Ignition coil internal circuitry

THE FIX

- Replaced the ignition coil with new one

(8) SUDDENLY THE VEHICLE LOSES POWER THEN PICKS UP AGAIN

- Vehicle Make: Land Rover
- Model: Range Rover Sport (SDV8)
- Year: 2012 - 2021
- Engine Code: 448DT
- Steering Wheel: Right Hand Drive
- Transmission: Automatic

PROBLEM DESCRIPTION BY VEHICLE DRIVER

The vehic le will be driving ok but then it just loses the power, then it picks up again and continues driving ok.

SYMPTOMS

- Loses power while driving then picks up again

HOW FAULT FINDING WAS DONE

- Test drove the vehicle to verify the symptoms
- Scanned the engine control module with the diagnostic scanner
- Trouble codes retrieved:
    - P0191 Fuel rail pressure sensor A circuit range/performance,
    - P0087 Fuel rail / system pressure too low

- Disconnected the connector from the fuel rail pressure sensor and tested for ground, voltage and signal wires (OK),hence the sensor is faulty.

WHAT CAUSED THE FAULT

- Fuel rail pressure sensor malfunctioned

THE FIX

- Replaced the fuel rail pressure sensor with new one

(9) MISFIRING

- Vehicle Make: Nissan
- Model: Note
- Year: 2006-2013
- Engine Code: CR14DE
- Steering Wheel: Right Hand Drive
- Transmission: Automatic

PROBLEM DESCRIPTION BY VEHICLE DRIVER

The car is misfiring and cannot idle properly

SYMPTOMS

- Engine shaking at idle

HOW FAULT FINDING WAS DONE

- Tested for 12V power supply, ground and signal in the wires to the ignition coils (ok)

- Removed spark plugs to inspect them. Spark plug 2 and 3 had larger spark plug gap compared to 1 and 4 hence the cause of misfiring.

WHAT CAUSED THE FAULT

- Spark plugs tear and wear

THE FIX

- Replaced all the four spark plugs with new ones

(10) THE CAR SWITCHES OFF WHEN YOU REV BEYOND 2000 RPM

- Vehicle Make: Ford
- Model: Focus
- Year: 1998-2005
- Engin e Code: FYDA
- Steering Wheel: Right Hand Drive
- Transmission: Automatic

PROBLEM DESCRIPTION BY VEHICLE DRIVER

When you press the accelerator pedal to increase the RPM the vehicle will switch off once RPM starts going beyond 2000 RPM.

SYMPTOMS

- Changes gears only up to Gear 2
- Switches OFF when accelerating beyond 2000 RPM but if youdrive slowly it will not switch OFF

HOW FAULT FINDING WAS DONE

- Scanned the car and retrieve trouble codes:
  - P0130: oxygen sensor heater circuit malfunction Bank 1 sensor 1,
  - P1131: lack of upstream heated oxygen sensor switch sensor indicates lean Bank 1
- Tested fuel pressure at fuel pump outlet and found very low pressure

WHAT CAUSED THE FAULT

- Fuel pump motor tear and wear

THE FIX

- Replaced the fuel pump motor with new one

OPTIONAL FIX

- Replace the complete fuel pump housing unit

(11) DELAYS ACCELERATION FROM TAKE OFF

- Vehicle Make: VW
- Model: Golf 5 TSI
- Year: 2007-2008
- Engine Code: CAXA
- Steering Wheel: Right Hand Drive
- Transmission: Automatic

PROBLEM DESCRIPTION BY VEHICLE DRIVER

The vehicle delays accelerating even when the accelerator pedal isdepressed to the end.

SYMPTOMS

- Delayed acceleration
- Black smoke if accelerator is depressed to the end

HOW FAULT FINDING WAS DONE

- Scanned the car computer and retrieved trouble code:
    - P 003A turbocharger or a supercharger boost pressure control a position exceeded learning limit
- Removed diverter valve and inspected it. Valve inside found stuck. Hence the cause of problem

WHAT CAUSED THE FAULT

- Tea r and Wear of the diverter valve

THE FIX

- Replaced turbocharger diverter valve with a new one

HOW FAULT FINDING WAS DONE

- Scanned the car computer and retrieved trouble code:

  P 003A turbocharger or a supercharger boost pressure control a position exceeded learning limit

- Removed diverter valve and inspected it. Valve inside found stuck. Hence the cause of problem

WHAT CAUSED THE FAULT

Tear and Wear of the diverter valve

THE FIX

- Rep laced turbocharger diverter valve with a new one

(12) START AND THEN IT STOPS IMMEDIATELY

- Vehicle Make: Mercedes-Benz
- Model: C-CLASS (203) C180K
- Year: 2002-2008
- Engine Code: 271. 946

- Steering Wheel: Right Hand Drive
- Transmission: Automatic

PROBLEM DESCRIPTION BY VEHICLE DRIVER

When you crank the car, it starts very well but will immediately switch OFF

SYMPTOMS

- Car starts then it switches off immediately or just after one second

HOW FAULT FINDING WAS DONE

- Tested fuel pressure (ok) Removed the downstream oxygen
- sensor from the exhaust pipe then started the car, car started without switching OFF. Hence the exhaust blocked on the catalytic converter after the down stream oxygen sensor.

WHAT CAUSED THE FAULT

Broken catalytic co n ve rte r pa rti cles causing block

THE FIX

- U nblocked the exhaust

OPTIONAL FIX

- Replace the catalytic converter with another one

(13) CRANKS BUT IT DOES NOT START

- Vehicle Make: Chevrolet
- Model: Optra
- Year: 2004-2010
- Engine Code: F16D3
- Steering Wheel: Right Hand Drive
- Transmission: Manual

PROBLEM DESCRIPTION BY VEHICLE DRIVER

The car was starting just ok before parking it. After parking it and came back to START it, the car refused to start.

SYMPTOMS

- Cran ks but does not start
- Dashboard lights come ON as normal
- Fuel pump can be heard running when ignition switch is ON then goes OFF after a second

HO W FAULT FINDING WAS DONE

- Sca nned the ECM and the trouble code retrieved were:
  - P0335, P0337
- Tested signal produced by camshaft position sensor while someone is cranking the car (ok)
- Tested signal produced by crankshaft position sensor while someone cranking the car (not ok)

WHAT CAUSED THE FAULT

Crankshaft position sensor malfunction

THE FIX

- Replaced the crankshaft position sensor

(14) ENGINE COOLANT TEMP SENSOR DTC UNABLE TO BE CLEARED

- Vehicle Make: VW
- Model: Polo TSI (6R1)
- Year: 2009-2014
- Engine Code: CBZ
- Steering Wheel: Right Hand Drive
- Transmission: Automatic

PROBLEM DESCRIPTION BY VEHICLE DRIVER

Check engine light appeared and stayed permanently on the dashboard. It does not go away after car has started. The car is still performing ok only that I don't like seeing the 'check engine' light appearing permanently on the dashboard.

SYMPTOMS

- Check engine light appearing permanently on the dash board
- . Car drives without problems

HOW FAULT FINDING WAS DONE

- Scanned the ECM and the trouble code retrieved was:

- P0117: Engine coolant temperature sensor (ECT) low input

- Tested the wires on the ECT sensor (OK), Hence the ECT is faulty

WHAT CAUSED THE FAULT

Tear and wear of the ECT sensor

THE FIX

- Replaced the engine coolant temperature sensor with new one

(15) ENGINE SHAKING VERY BADLY WHEN IDLING

- Vehicle Make: Chevrolet
- Model: Optra
- Year: 2004-2010
- En g i ne Code: F16D3
- Ste e r i ng Wheel: Right Hand Drive
- Tra n smission: Manual

PROBLEM DESCRIPTION BY VEHICLE DRIVER

After the previous mechanic worked on the car that's when the shaking of the engine started

SYMPTOMS

- Mi sfiring
- Engine shaking at idle
- Check engine light stays ON while car is idling

HOW FAULT FINDING WAS DONE

- Tested for 12V power supply to the ignition coils (OK)
- Tested for ground to the ignition coils (OK)

- Ignition coil had connector removed by previous mechanic and swapped its wires without noticing, hence the cause of ignition coil for cylinder number 1 not firing.

WHAT CAUSED THE FAULT

Wires on the connector to the ignition coil swapped by previous mechanic

THE FIX

- Connected the two wires to the ignition coil the correct way

(16) IDLING TOO HIGH OVER 2000 RPM

- Vehicle Make: Toyota
- Model: Vios
- Year: 2001-2006
- Engine Code: 1NZ
- Steering Wheel: Right Hand Drive
- Transmission: Manual

PROBLEM DESCRIPTION BY VEHICLE DRIVER

After cleaning the throttle body with carburetor cleaner that's when the problem of high idling started

SYMPTOMS

- Fuel consumption too high
- RPM at 2000 or nearly over 2000 during idling

HOW FAULT FINDING WAS DONE

- Checked the adjustment of throttle cable (OK) Unplugged connector from ISCV (idle speed control valve) but still RPM was 2000 at idle, hence the cause of the problem

WHAT CAUSED THE FAULT

- ISCV malfunction

THE FIX

- Replaced the ISCV idle speed control valve

(17) STARTS FOR ABOUT A SECOND AND THEN SWITCHES OFF

- Vehicle Make: VW
- Model: GOLF V (GTI)
- Year: 2004-2008
- Engine Code: AXX
- Steering Wheel: Right Hand Drive
- Transmission: Automatic

PROBLEM DESCRIPTION BY VEHICLE DRIVER

While driving the car, the timing belt broke. During towing to the garage using towing bar hooked in front, smoke started coming out from the gearbox area. At the garage, the mechanic replaced the cylinder head with another good one, but after that car refused to start.

SYMPTOMS

- Check engine light warning symbol not appearing on the dashboard when ignition switched to ON
- Fuel pump runs as normal and then switches off when key is ON
- Key w arning symbol comes ON and goes OFF as normal when ignition switch turned to ON.

HOW FAULT FINDING WAS DONE

- Scannedthevehicle ECM and retrieved trouble code:
  - U0001: CAN bus high.

  (This trouble refused to be cleared)

- Unplug the connector to the transmission control module then switched ignition switch to ON, then check engine light appeared on the dashboard, cranked the engine directly by jumping a wire from positive battery terminal to starter solenoid and car started and maintained the idling. If connector is plugged back to the transmission control module then check engine light will not appear on dash board and car will not start, hence TCM is the cause of problem.

WHAT CAUSED THE FAULT

- Transmission control module (TCM) burnt during wrong procedure of towing the car

THE FIX

- Replaced complete set of gearbox and TCM

OPTIONAL FIX

- Repair gearbox and TCM

(18) CRANKS BUT NO START

- Vehicle Make: Toyota
- Model: Hilux
- Year: 1989-1993
- Engine Code: 4 Y
- Steering Wheel: Right Hand Drive
- Transmission: Manual

PROBLEM DESCRIPTION BY VEHICLE DRIVER

While driving along the way, vehicle just stopped. I parked vehicle on the side of the road and waited for about an hour and tried to start and the vehicle started. Drove the vehicle and arrived home and parked the vehicle. The following morning tried to start the vehicle but this time it refused to start. Three weeks passed but still the car would not start.

SYMPTOMS

- No spark
- Can crank but no start

HOW FAULT FINDING WAS DONE

- Checked the spark from all the 4 high tension leads (no spark) Tested for power supply to the distributor (ok)
- Checked the condition of the distributor cap contact
- brushes, they were found worn out hence the cause of the problem

WHAT CAUSED THE FAULT

Distributor cap worn out contact brushes

THE FIX

- Replaced the distributor cap with a new one

(19) CRANKS BUT DOESN'T START

- Vehicle Make: Toyota
- Model: Gaia
- Year: 1998 - 2004
- Engine Code: 1AZ-FSE
- S teering Wheel: Right Hand Drive
- Transmission: Automatic

PROBLEM DESCRIPTION BY VEHICLE DRIVER

When the ignition switch is turned to ON, the check engine light does not appear on the dashboard but all the other dashboard lights come ON as normal. The dashboard lights appear in very faint colour but sometimes they will be bright the normal way. The radiator fan will start running by itself when the ignition switch is turned to ON.

SYMPTOMS

- Cranks but not starting
- When ignition switch is turned to ON, all the dashboard lights will appear as normal but in faint color and check engine light will not appear.
- When ignition switch is turned OFF, all the dashboard lights will go OFF but then the check engine light will appear in very faint color then radiator fan will start running.

HOW FAULT FINDING WAS DONE

Checked all the earth points around the engine, one earth point was found loosely bolted onto the engine, hence the cause of the problem.

WHAT CAUSED THE FAULT

- Loose bolt on the earth point on the engine

THE FIX

- Retightened the loose earth point according to torque specifications

(20) STARTS BUT CANNOT REV UP

- Vehi cle Make: Toyota
- Model: Gaia
- Year: 1998 - 2004
- Engine Code: 1AZ-FSE
- Steering Wheel: Right Hand Drive
- Transmission: Automatic

PROBLEM DESCRIPTION BY VEHICLE DRIVER

While driving, the car just lost power and acceleration

S Y M P T O M S

- Starts and stays on idle
- Only idling even after depressing accelerator pedal
- RPM does not change, stays permanently at idle
- Check engine light permanently appearing on dashboard even when the car is idling

HOW FAULT FINDING WAS DONE

- Checked the fuse for the throttle body on the fuse box in the engine bay. The fuse was found blown, hence the cause of the prob lem. The fuse box is located in the engine bay, right side, the area near the alternator.

WHAT CAUSED THE FAULT

Over current drawn through the fuse due to aging of some components

THE FIX

- Replaced the blown fuse with new one and checked if it blew up again but did not blow up

(21) HARD START

- Vehicle Make: VW
- Model: Polo (9N1/9N2/9N3)
- Year: 2005-2009
- Engine Code: BKY

- Steering Wheel: Right Hand Drive
- Transmission: Automatic

PROBLEM DESCRIPTION BY VEHICLE DRIVER

I bought this car as a second hand car with this hard start problem existing on it. I took the car to different mechanics so they can fix the hard start problem but the problem is not getting fixed. What it does is that in the morning, the car will start at one kick. After you have driven the car for some time and you switch off and try to start after lets 10 minutes it will have a hard start. But if you switch off and t ry to start immediately it will start at one kick without hard start. As the hard start happens, when starting you need to keep the key at START position for about 7 seconds as it cranks then it eventually starts. The car moves very fine after starting.

SYMPTOMS

- In the morning, first time starting the car it starts at one kick
- During hard start, key needs to be at START position for about 7 seconds as it cranks before it eventually starts
- If you t r y to start immediately after switching off the car it will start a t one kick
- There are no warning lights appearing on the dash board

HOW FAULT FINDING WAS DONE

- Diagnosed all the systems on the car. No trouble codes were stored
- Inspected the spark plugs, spark plugs appearance indicated some slight rich fuel mixture
- Inspected the fuel injectors. All the four injectors were noticed that they were half way out from there ports. Thorough inspection on the injectors showed that bigger O-ring seals were used on the injectors causing the injectors not to be fully inserted into their ports. The injectors not fully inserted into their ports caused the late arrival of fuel into the cylinders, hence the cause of the problem

WHAT CAUSED THE FAULT

- Injectors not fully inserted into their ports because of bigger O-ring seals used

THE FIX

- Replaced all the four injectors with good second hand injectors that had the correct O-ring seals around them such that all the four injectors were fully inserted into their ports

(22) TOO MUCH ERRATIC IDLING

- Vehicle Make: VW
- Model: GOLF V (GTI)
- Year: 2006-2008
- Engine Code: BWA
- Steering Wheel: Right Hand Drive
- Transmission: Automatic

PROBLEM DESCRIPTION BY VEHICLE DRIVER

The car just started misfiring so badly from nowhere. The engine will be shaking too much while on idol.

S Y M P T O M S

- Engine shaking at idle
- Check engine light not going off after starting car
- Erratic idling with some rattling noise from engine

HOW FAULT FINDING WAS DONE

- Scanned the ECM and retrieved the following trouble codes:
    - P0444 evaporative emission system purge control valve circuit open

- P0033 turbo charger bypass valve control circuit
- P0243 turbo/supercharger waste gate solenoid (A): open or short circuit to ground
- P1355 cylinder 1 ignition circuit: open circuit
- P0300 random or multiple cylinder misfire detected
- P0301 cylinder 1 misfire detected

- When trouble codes are cleared they will come back immediately
- Checked fuse F9 and was found blown up, hence the cause of P0444, P0033 and P0243.Fuse F9 is supplying turbocharger waste gate solenoid, turbocharger bypass valve and evaporative emiss ion purge control valve.
- Tested 12V power supply to ignition coil 1 (OK)
- Tested for ground to ignition coil 1 (OK)
- Tested signal to ignition coil 1 (OK)
- Inspected sparkplug 1 (OK)
- Swapped ignition coil 1 with coil 2 and started the car then scann ed the car with diagnostic scanner. Cylinder 2 reported misfiring on the scanner, hence this ignition coil is faulty.

WHAT CAUSED THE FAULT

- Fuse F9 blown up due to overcurrent
- Ignition coil malfunctioned internally

THE FIX

- Installed new fuse F9 with same amp rating as the blown one and observed if it would blow up but did not blow up after car started

- Replaced malfunctioned ignition coil with new one

(23) CAR IS OVERHEATING

- Vehicle Make: VW
- Model: PASSAT
- Year: 2005-2009
- Engine Code: BLR
- Steering Wheel: Right Hand Drive
- Transmission; Automatic

PROBLEM DESCRIPTION BY VEHICLE DRIVER

Temperature guage rising going closer to the red zone. The water started boiling in the water bottle.

SYMPTOMS

- Temperature gauge rising beyond normal
- Water boiling in the water bottle

HOW FAULT FINDING WAS DONE

- Tested for constant power supply to the fan control mode in the red wire, constant power supply was missing.
- Checked the fuse protecting the radiator fan. Fuse was found blown up.

- Replaced the fuse then fun started running continuously. To stop t he fun, battery terminal should be removed. The fuse that
was found blown up indicated that the fan control module was faulty internally

WHAT CAUSED THE FAULT

- Fan control module internal short circuits

THE FIX

- Replaced the big radiator fan and the fuse. (Fan control module and the big fan are on one assembly.

(24) OVER HEATING BUT RADIATOR FAN IS RUNNING OK

- Vehicle Make: BMW
- Mod el: 3 SERIES (E90/91/92/93) (323i)
- Year: 2005-2008
- Engine Code: N52 B25A
- Steeri ng Wheel: Right Hand Drive
- Transmission: Automatic

PROBLEM DESCRIPTION BY VEHICLE DRIVER

The water in the water bottle boils and spills out. Then the low coolant warning light appears on the dashboard. The radiator fan running quite OK.

SYMPTOMS

- Overheating
- Water boils in water bottle
- Check engine light appearing
- Water coolant warning light appears on dashboard

HOW FAULT FINDING WAS DONE

- Scanned the car with diagnostic equipment and retrieved trouble code:
    - P1619 engine coolant thermostat control circuit: signal low
- Tested the 12volts power supply to the thermostat (OK)
- Tested the control wire to the thermostat (OK)
- All te sts proved to be OK, then thermostat must be replaced
- Thermostat is on the thermostat housing just under the electric water pump on the right bottom section of the engine

WHAT CAUSED THE FAULT

Tear and wear of the engine coolant thermostat

THE FIX

Replaced the entire engine coolant thermostat housing

(25) REV COUNT FLUCTUATING DURING IDLING

- Vehicle Make: Nissan
- Model: Tiida
- Year: 2007-2011
- Engine Code: HR16DE
- Steering Wheel: Right Hand Drive
- Transmission: Automatic

PROBLEM DESCRIPTION BY VEHICLE DRIVER

After the car was serviced, the rev count could not stabilize. At service they replaced the spark plugs. They first removed the throttle body to access the spark plugs.

SYMPTOMS

- RPM's unstable
- RPM's fluctuating between 1000 and 2000
- Fue l consumption too high

HOW FAULT FINDING WAS DONE

- Dia gnosed the computer box (ECU) with the diagnostic scanner. The trouble codes were talking about the throttle body

WHAT CAUSED THE FAULT

The removal of throttle body during service

THE FIX

- Did the throttle body relearn procedures using the diagnostic scanner

(26) HARD START

- Vehicle Make: VW
- Model: Golf GT (FSI)
- Year: 2005 -2008
- Engine code: BLR
- Steering Wheel: Right Hand Drive
- Trans mission: Automatic

PROBLEM DESCRIPTION BY VEHICLE DRIVER

I bought the car as second hand with this problem of hard start. The hard start is so much that sometimes it will even drain the battery.

Sometimes it can just start at one kick but sometimes I will have to crank up to four times for the car to start. After the car has started it will move very fine. There is no specific time when the hard start happens. It happens so often at any time.

SYMPTOMS

- Check engine warning light appearing on the dash board
- Sometimes you need to crank up to four times for car to start
- Sometimes it can start on the first crank
- The hard start happens more often than it starts on the first crank

HOW FAULT FINDING WAS DONE

- Diagnonesed the computer box (ECU) with diagnostic scanner.

    - P 0341 – Camshaft position sensor (G40): Implausible signal

- Check ed the camshaft timing in relation to the crankshaft TDC. The camshaft timing was found out of timing, hence the cause of the hard start.

WHAT CAUSED THE FAULT

Camshaft timing not correct (misaligned)

THE FIX

- Re-timed the camshaft timing in relation to the crankshaft TDC

(27) MISFIRING

- Vehicle Make: HONDA
- Mode l: FIT
- Year: 2001-2008
- Engine code: L13A
- Steering Wheel: Right Hand Drive
- Transmission: Automatic

PROBLEM DESCRIPTION BY VEHICLE DRIVER

The misfiring started just from nowhere.

SYMPTOMS

- Engine hesitating during idle

HOW FAULT FINDING WAS DONE

- Tested for power supply to the ignition coils on bank two. Power supply on the yellow wire that supply the four ignition coils on bank two was missing, hence the cause of misfiring.

WHAT CAUSED THE FAULT

- The wire supplying power to the ignition coins on bank two is broken somewhere.

THE FIX

- Tap power supply from a reliable wire on the fuse box to supply the ignitions on the bank two when ignition switch is turne d ON.

(28) NOT STARTING ONLY CRANKING

- Vehicle Make: TOYOTA
- Model: HARRIER
- Year: 2001-2005
- Engine Code: 1MZ
- Steering Wheel: Right Hand Drive
- Transmission: Automatic

P ROBLEM DESCRIPTION BY VEHICLE DRIVER

I was driving the car then it slowly lost the power and eventually switched OFF. We towed the car and brought it home. One mechanic came to check it but did not manage to fix it.

SYMPTOMS

- Only cranks Not starting All lights on the dashboard come ON
- as normal when ignition switch is ON
-

HOW FAULT FINDING WAS DONE

W

- Diagnosed the car with the diagnostic equipment. The diagn ostic equipment reported that there were no fault codes

- Set the diagnostic equipment on ‘LIVE DATA’ and cranked the car so I could see on the live data if the injectors were responding to the cranking. The live data showed that the injectors were responding to the cranking.

- Tested for spark at the engine on one of the ignition coils (OK)

- Activated the fuel pump relay using the diagnostic equipment, the re lay clicking sound was head but the fuel pump did not run.

- Removed outlet pipe from the fuel pump then connected fuel pressure guage on it and cranked the car to observe if pressure guage will respond but pressure guage did not respond. Fuel pump is not running hence the cause of the problem.

WHAT CAUSED THE FAULT

- Fuel pump tear and wear

THE FIX

- Replaced the fuel pump motor only

(29) WATER IS BOILING IN THE WATER BOTTLE

- Vehicle Make: VW
- Model: Polo (9N1/9N2/9N3)
- Year: 2002-2006
- Engi ne Code: BBY
- Steering Wheel: Right Hand Drive
- Transmission: Automatic

PROB L EM DESCRIPTION BY VEHICLE DRIVER

Temperature gauge is rising approaching the red zone maximum. The radiator fan does not run even when the temperature guage is rising beyond normal. Very hot steam is coming out from the water bottle.

SYMPTOMS

- Low coolant warning light appears on the dashboard
- Steam coming out from the engine Bay
- Coolant temperature guage rising to maximum
- Radiator Fan not running

HOW FAULT FINDING WAS DONE

- Inspe cted the engine Bay for water leaks
- Inspected all the water hoses
- Inspected the thermostat housing. Thermostat housing was found broken on one of its outlets, hence water leaking through it.
- When the cooling system is empty of water or coolant, the radiator fan will not run as there would be no water or coolant to heat up the radiator thermal switch

WHAT CAUSED THE FAULT

- Thermostat housing broken due to tear and wear

THE FIX

- Replaced the thermostat housing

(30) CRANKS BUT NO START

- Vehicle Make: Audi
- Model: A4 (8Ex)
- Year: 2001-2008
- Engine Code: ALT
- Steering Wheel: Right Hand Drive
- Transmission: Automatic

PROBLEM DESCRIPTION BY VEHICLE DRIVER

- Before replacing the gearbox, the car was starting. After replacing the gearbox, the car only cranks but not starting

SYMPTOMS

- Dashboard turns ON as normal
- Cranks but does not start

H OW FAULT FINDING WAS DONE

- Tested for power supply to all the four ignition coils on their connectors (ok)
- Tested for the earth on all the connectors. Each connector has two earths.one earth was missing on all the connectors, hence the cause of the problem

WHAT CAUSED THE FAULT

- One earth to the connectors was not attached to its earth point on the car

SYMPTOMS

- Resto red the missing earth to all the four connectors

(31) S PARK PLUG CERAMIC MATERIAL HAS OIL AROUND IT

- Vehicle Make: VW
- Model: POLO
- Year: 2006 - 2009
- Engine Code: CLP
- Steering Wheel: Right Hand Drive
- Transmission: Manual

PROB L EM DESCRIPTION BY VEHICLE DRIVER

Engine oil is found around the spark plug ceramic material on all the four spark plugs. When the oil is wiped from the spark plugs it will appear again just after a few days.

SYMPTOMS

- Car drives normally
- Engine oil is noticed in the spark plug holes

HOW FAULT FINDING WAS DONE

- Removed the ignition coils from all the four plugs while car is OFF. (Engine oil is seen around the ignition coil metal sheath)
- Some amount of engine oil is seen in the spark plug holes
- Removed all the four spark plugs (Engine oil is seen around the spark plug ceramic material)
- Hence the engine oil leaks into the spark plug holes through the o-ring/dichtring. (The dichtring is the seal that is placed in between the top of the spark plug holes and the surface of the camshaft box)

WHAT CAUSED THE FAULT

- O-ring/Dischtring worn out

T HE FIX

- Replaced the o-ring/dichtring with new one by first removing the camshaft box

(32) MISFIRING AND ENGINE SHAKING

- Vehicle Make: VW
- Model: GOLF 6 TSI
- Year: 2008-2013
- Engin e Code: CAXA
- Steering Wheel: Right Hand Drive
- Transmission: Automatic

PROB LEM DESCRIPTION BY VEHICLE DRIVER

The vehicle started misfiring suddenly and the engine will be shaking. It has also lost the power such that it will delay to accelerate. The gear shifting is also struggling such that sometimes the maxim um gear shift it will go is gear 4.Before this misfiring started the car was driving just fine even the gear shifting was also good all the way to gear 7.

SYMPTOMS

- Misfiring
- Engine shaking
- Delays to accelerate
- Shifting of gears is not smooth

HOW FAULT FINDING WAS DONE

- Diagnosed the car using diagnostic equipment and retrieved trouble codes:
    - P0300 random/multiple cylinder misfire detected,
    - P0301 cylinder 1 misfire detected,
    - P0303 cylinder 3 misfire detected
    - P130A hide cylinder
- Remo ved spark plug from cylinder 1 to inspect it and test the cylinder for compression. The spark plug electrode and the ceram ic coating around the electrode was found broken. The compression test was reading less than 4 bars. According to the manual for this car, compression is supposed to read between 7 to 15 bars. Hence the misfire in cylinder 1 is caused by the low compression and the broken spark plug electrode.

- Removed spark plug from cylinder 3 to inspect it and test the cylinder for compression. Spark plug was found wet with a lot of engine oil around the electrode. The compression test was readin g 5.9 bars. According to the manual of this car, compression is supposed to read …between 7 to 15 bars. Hence the misfire in cylinder 3 is caused by low compression and the engine oil around the spark plug electrode.

- Removed the cylinder head from the engine to inspect the valve s . One exhaust valve for cylinder 1 was found chipped off from its edge. Hence the cause of low compression in cylinder 1.It is assumed that the pieces that broke off from the spark plug in cylinder 1 are the ones that chipped off a piece from the exhaust valve. The rest of all the other valves were found to be in good condition without any leaks on them after testing them for leaks. Hence the low compression in cylinder 3 will be

  attributed tothe piston rings.

WHAT CAUSED THE FAULT

- Low compression in cylinder 1 and 3

THE FIX

- Rebuild the entire engine with new replacement parts

OPT IONAL FIX

- Replace engine with another engine

(33) MILKY SLUDGE ON THE ENGINE OIL FILLER CUP

- Vehicle Make: Mercedes Benz
- Model: E200
- Year: 2003 - 2009

- Engine Code: M271.956
- Steer i ng Wheel: Right Hand Drive
- Transmission: Automatic

PROBLEM DESCRIPTION BY VEHICLE DRIVER

The coolant hose to the thermostat housing keeps on being disconnected during driving to the point that the car started overheating. Then one day I checked the engine oil filler cup and saw the milky sludge on it and inside the engine. That is when I realized that engine oil had mixed with the coolant

SYMPTOMS

- Over heating
- Milky sludge on the engine oil filler cup and inside the engine looking through the engine oil filler hole
- Engi ne oil in the engine sump appearing with small amounts of milky looking sludge in it

HOW FAULT FINDING WAS DONE

- Inspected the engine oil appearance on the engine oil filler cup. The e ngine oil appeared milky on the filler cup and inside the engine through the engine oil filler hole. Milky sludge on engine oil filler cup indicates that engine oil had mixed with coolant, hence the cause of the problem

- Inspected the engine oil dip stick. Engine oil on the dip stick did n o t look milky since the engine oil coolant mixture was not too much in the engine oil sump

WHAT CAUSED THE FAULT

- Overheating of the engine

THE FIX

- Removed the cylinder head and had it resurfaced/skimmed, replaced cylinder head gasket with new one and reinstalled the cylinder head.

OPTIONAL FIX

- Replace cylinder head or Replace complete engine

(34) CRANKS BUT NO START

HOW I

- Vehicle Make: Toyota
- Model: GAIA
- Year: 1998-2004
- Engin e Code: 3S-FE
- Steering Wheel: Right Hand Drive
- Transmission: Automatic

PROBL EM DESCRIPTION BY VEHICLE DRIVER

The car was bought as a non-runner. The previous mechanic did the engine overhaul but car is only cranking and not starting.

SYMPTOMS

- Dashboard light come ON as normal
- Cranks but no start

HOW FAULT FINDING WAS DONE

- Teste d for sparked (ok)
- Tested for 12V power supply to the injectors (No power), hence the cause of the problem

WHAT CAUSED THE FAULT

- Powersupplywiretoinjectorsbrokensomewhereinthe wiring harness

TH E FIX

- Restored 12V power supply to the injectors

(35) CRANKS BUT NO START

- Vehicle Make: Volkswagen
- Model: Golf V (GTI)
- Year: 2006-2008
- Engine Code: BWA
- Steering Wheel: Right Hand Drive
- Transmission: Automatic

PROBLEM DESCRIPTION BY VEHICLE DRIVER

I arrived home driving the car. I let the car idle for some time then started reversing so I can repark nicely. As I reversed, the car suddenly lost power and eventually switched OFF. Tried to start it but could only crank not starting.

SYMPTOMS

- Dashboard lightcomeON
- Cranksbut no start
- Fuelpump is heard running onignitionswitchON

HOW FAULT FINDING WAS DONE

- Diagnosed the car and retrieved trouble code:
    - P0087:fuelrail/systempressure too low

- Teste d for fuel pressure on the fuel pump outlet with the pressure guage. Pressure reading below 1 bar as someone is cranking the car. Fuel pressure for this car is supposed to be 4 to 6 bars, hence fuel pump is condemned.

WHAT CAUSED THE FAULT

- Fuel pump motor tear and wear

THE FIX

- Replaced the complete fuel pump unit with new one.

CHAPTER 2

TRANSMISSION PROBLEMS

(36) SPEED NOT GOING BEYOND 80 KM PER HOUR

- Vehicle Make: Mercedes
- Model: C- CLASS (203) C200K
- Year: 2000-2003
- Engine Code: 111. 955
- Steering Wheel: Right Hand Drive
- Trans mission: Automatic

PROBLEM DESCRIPTION BY VEHICLE DRIVER

Repairs were recently done in the engine bay and after that, the problem started. The speed cannot go beyond 80 km per hour. You need to depress the accelerator pedal beyond 3000 RPM for the speed to reach 80 km per hour.

SYMPTOMS

- Gear changelimited
- Gear leverstuckin Park, canonly shift toother position if you press theemergency button nextto Gear lever

H OW FAULT FINDING WAS DONE

- Battery negative terminal disconnected and terminal connector plug disconnected from the ABS control module. Reconnected battery negative terminal while ABS still disconnected

- Tested for 12 volts constant power supply on pin 32 on the terminal connector plug (ok)

- Tested for 12 volt constant power supply on pin 1 on the terminal connector plug (no power)

- Tested for 12 volt ignition switch ON on pin 4 on the terminal connector plug (no power)

WHAT CAUSED THE FAULT

Fuse F59 and Fuse F56 missing on their place on the fuse box.F59 and F56 supply 12V power to Pin 1 and Pin 4 respectively. Previous mechanic forgot to fit the fuses F59 and F56 back in their places after working around the engine bay.

THE FIX

- Fitted fuse F59 and fuse F56 of the correct amp rating

(37) GEARS NOT CHANGING

- Vehicle Make: VW
- Model: Polo
- Year: 2005-2009
- Engine code: BKY
- Steering: Right Hand Drive
- Transmission: Automatic

PROBLEM DESCRIPTION BY VEHICLE DRIVER

The gears suddenly just stopped changing. The PRND123 lights are not showing on the dash board.

SYMPTOMS

- PRND123 not showing on the dashboard
- Forw ard driving only in a single gear
- Check engine light showing on the dashboard

HOW FAULT FINDING WAS DONE

- Scanned the vehicle with diagnostic scanner. The DTC in the computer box (ECU) talks about "lost communication with transmission control module

- The diagnostic scanner fails to communicate with tran smission control module

- Tested for constant 12 volts positive power supply to transmission control module (MISSING)

- Tested for ignition switch ON 12 volts positive power supply to transmission control module (OK)

- Inspected fuse F13.Fuse F13 was found blown. Fuse F13 is supplying constant 12 volts positive to the transmission control module.F13 is a 5A fuse.

- Replaced the blown fuse with a good one of the same amp rating and reconnected the TCM to see if the fuse will blow up again. Immediately the fuse blew up. Hence the fault is the TCM.

- Opened the TCM to inspect for burnt marks. TCM was found burnt inside.

WHAT CAUSED THE FAULT

- TCM internal circuitry short circuited

THE FIX

- Replaced the TCM with the working one.

(38) GEARS NOT CHANGING

- Vehicle Make: Audi
- Model: A3
- Year: 2005-2012
- Engine code: BSE
- Steering: Right Hand Driving
- Transmission: Automatic

PROBLEM DESCRIPTION BY VEHICLE DRIVER

I was driving then suddenly heard a bang sound kind of like a very harsh gear shift. I stopped the car on the side of the road and switched i t OFF. Then STARTED it again but the moment I moved the gear lever to the R (Reverse) the car jerked with a bang sound. Then shifted the gear to N (neutral) there was neither jerking nor sound. The moment I shifted the gear to D (drive) again the car jerked with a bang sound then gear '3' appeared on the dashboard. As I mov ed the car it did not change to any gear but only drove in gear 3.

SYMPTOMS

- Harsh reverse (R) gear engaging with a bang sound
- Harsh drive (D) gear engaging with a bang sound
- Forward drives only in gear 3 and does not change into any other gear
- The PRNDS letters are all highlighted/shaded

HOW FAULT FINDING WAS DONE

- Diagnosed engine control module and the transmission control module with a diagnostic scanner and retrieved trouble codes: -
    - P 1652 please check DTC memory of: transmission ECU
    - Solenoid 3 N90 sp blocked or without voltage

- Inspected terminal connector at the valve body on the gearbox. The pin 7 on the connector was found pulled out from its terminal connector housing, hence the cause of the problem.

WHAT CAUSED THE PROBLEM

- Tear and wear of the terminal connector

THE FIX

- Fixed the pulled out pin 7 back on its place on the terminal connector.

OP TIONAL FIX

- Replace the terminal connector

(39) NOT ENGANGING GEARS

- Vehicle Make: VW
- Model: Polo
- Year: 2010
- Engine code: CAV
- Steering: Right Hand Drive
- Transmission: Automatic

PROBLEM DESCRIPTION BY VEHICLE DRIVE

I noticed while the car was driving that the "D" started blinking. I stopped a nd opened the bonnet then saw oil leaking out from the top of the mechatronic unit. I managed to drive the car home and parked it. The following morning after STARTING the car it refused to reverse after moving the gear selector to R (reverse). Then selected D (Drive) but still did not move.

SYMPTOMS

- Car STARTS but cannot move either forward or Reverse Rever se/Drive does not engage
-
- Oil leaking out from the mechatronics breather cap while car is idling

- Smoke like substance coming out from the mechatronics breath er cap while car is idling

HOW FAULT FINDING WAS DONE

Diagnosed the transmission control module with diagnostic scanner and retrieved trouble codes: - P17BF hydraulic pump: play protection :

- P17BF hydraulic pump: Play protection (upper limit exceeded)
- P17BF hydraulic pump: Play protection (lower limit exceeded)
- P189C function restriction: due to insufficient pressure build-up (Insufficient pressure build up in the mechatronics will require a rebuild of the mechatronics unit.)

WHAT CAUSED THE FAULT

- Tear and wear of the mechatronics component

THE FIX

Rebuilt the mechatronics unit with the following replacement parts :

- Valve body
- Gaskets

- O-rings
- Seals
- Accu m ulator
- Hydraulic pump motor
- Hydraulic Pump
- Mechatronics oil/green power steering oil

(Though basic settings were not performed, the car started changing gears just fine)

OPTIONAL FIX

- Replace complete mechatronics unit

(40) BANG SOUND WHEN REVERSE OR DRIVE IS SELECTED

- Vehicle make VW
- Mode l: Golf V
- Year: 2005-2008
- Engine code: BLX
- Steering: Right Hand Drive
- Transmission: Automatic

PROBLEM DESCRIPTION BY VEHICLE DRIVER

Vehicle had an engine problem before the engine was replaced. After the engine replacement that is when this problem started appearing. What it is doing is that it makes sound from gearbox when R or D is selected and while driving it looks like it is moving only in a single gear and not changing to any other gears. Before engine replacement everything was just fine.

SYMPTOMS

- PRNDS not showing on the dashboard when key is turned to ON
- Very loud bang sound from the gearbox when gear lever is moved to R (reverse) or D (drive)

HOW FAULT FINDING WAS DONE

- Diagnosed the transmission control module with a diagnostic scanner and retrieved trouble codes:
    - 01312 data bus drive
- Then diagnosed the engine control module and the retrieved codes:
    - U0001 powertrain system database faulty
- Diag nosed the dashboard [instrument cluster] and retrieved trouble codes:
    - 01309 steering assistance control unit
    - 01315 gearbox control unit
    - 01316 abs control unit
    - 01321 control unit for airbag
- The diagnostic scanner failed to scan the following control modules:
    - Steering assistance control module
    - ABS control module

- Airbag control module

- Disconnected the terminal connector from the transmission control module. Then switched ignition to ON and measured the voltage on the two CAN bus wires on the TCM connector. One wire was measuring voltage of 12 volts when measured against the ground. The other wire was measuring 0.16 volts. Hence there is a short to voltage in the CAN bus wires.

- Located and disconnected all the control modules in the powertrain databus according to the wiring diagram. The disconnected modules where: transmission control module, engine control module, ABS control module, gear shift control module, airbag control module and power steering control module. These modules were disconnected one after the other. Each time the the module was disconnected, the ignition switch was turned to ON and the voltage remeasured in the CAN bus at the TCM terminal connector. The last module to be disco nnected was the power steering control module. During the disconnecting of the power steering control module, the wires on this connector were found to be burnt. After power steering cont rol module connector was disconnected, the PRNDS was still ap pe ar ing hi gh li gh te d / sh a de d . T control module and the PRNDS started appearing normally with out being highlighted/shaded, hence engine control module is also faulty.

- Replaced the engine control module with a good working one and reconnected all the control modules except the power steeri ng control module and rescanned all the reconnected modules. The diagnostic scanner was now able to communicate with all the control modules.

WHAT CAUSED THE PROBLEM

- The burnt wires on the power steering control module connector. (Power steering control module connector is located somewhere between the battery and the left headlight, it is a small 6 pin connector.)

THE FIX

- Replaced the burnt wires on power steering control module connector
- (The power steering control module was left disconnected. When the power steering control module is reconnected the PRNDS would appear highlighted/shaded.)
- Repla ced the engine control module

O PTIONAL FIX

- Also include the replacement of the power steering control modu le

(41) CLUNKING SOUND WHEN CHANGING GEARS

- Vehicle make: Audi
- Model: A3
- Year: 2005-2012
- Engine code: BSE

- Transmission: Automatic
- Steer i ng: Right Hand Drive

PROBLEM DESCRIPTION BY VEHICLE DRIVER

There is a clunking sound that comes out from the gearbox when the vehicle changes gears in the upshift or downshift. The clunking sound is much worse when car is downshifting especially to gear 2 and gear 1.

SYMPTOMS

- Bang ing sound when R (reverse) is selected
- Banging sound when D (drive) is selected
- Clunking sound when car is upshifting or downshifting of the gears

HOW FAULT FINDING WAS DONE

- Inspected the center gearbox mounting. The dog born arm of the mounting was found damaged and loose on its bushing.

- Inspected the inner and outer CV joints, the inner CV joint of the right side was found damaged on its rubber boot and all the grease escaped.

WHAT CAUSED THE PROBLEM

- Worn out gearbox mount and inner CV joint

THE FIX

- Replaced the dog bone arm only of the mounting with new one
- Replaced the inner CV joint

OPTIONAL FIX

- Repl ace the entire center gearbox mount

(42) B ANGING SOUND WHEN THE REVERSE HOWO FRA DURLITV EFI INSD SIENLGE AE

- Vehicle Make: VW
- Model: Golf IV
- Year: 2000-2003
- Engine code: AVU
- Steering: Right Hand Drive
- Transmission: Automatic

PROBLEM DESCRIPTION BY VEHICLE DRIVER

The car was moving just fine then suddenly this problem appeared. Whenever I select the gear lever to reverse or drive there will be a very loud bang sound and the vehicle then jerks.

SYMPTOMS

- Very loud bang sound from the gearbox when gear lever is moved to R (reverse) or D (drive)
- Forward driving only in a single gear and not shifting to any other gear

HOW FAULT FINDING WAS DONE

- Diagn osed the transmission ECU with a diagnostic scanner and retrieved the trouble code:
  - 00297 gearbox speed sensor (G38) no signal
- Inspe cted the speed sensors that are on top of the gearbox. One speed sensor that is a two- pin sensor was found cut on one of its cables hence the cause of the problem

WHAT CAUSED THE PROBLEM

Broken cable to gearbox speed sensor

THE FIX

- Soldered the broken cable of the speed sensor

(43) GEARBOX BANGS WHEN UPSHIFTING OR DOWNSHIFTING

- Vehicle make: VW
- Model: Polo
- Year : 2005-2009
- Engine code: BKY
- Steering wheel: Right Hand Drive
- Tran smission: Automatic

PROBLEM DESCRIPTION BY VEHICLE DRIVER

This problem just started recently. What it is doing is that gear changing is hash and followed by a bang sound. This problem comes and goes. Sometimes the car will drive very nicely and changes gears nicely but then suddenly the problem will appear and the gears start banging when upshifting or down shifting. I have noticed that this is happening when the outside temperature is very hot for example during m idafternoon. And this problem will start the moment the traction control light appears on the dashboard.

SYMPTOMS

- Gear shifting upwards or downwards followed by a bang sound
- When the traction light appears on the dash board, the idling becomes slightly rough with the RPM needle gauge moving up and down slowly
- When the traction light does not appear on the dashboard then the car will drive just fine

HOW FAULT FINDING WAS DONE

- Diagnosed the engine computer box [ECU] but no trouble codes were stored
- Diagn osed the transmission control module and retrieved trouble codes:
    - 01166 engine torque signal no signal
    - 01166 engine torque signal outside Torrance
- Test drove the car during midafternoon hours till the problem resurf aced, then let the car idle as the traction light was appearing on the dashboard. As the car was idling I noticed that the idling had become rough. While car was on rough idle, all the four ignition coils were tested. Ignition coil number 2 was found not to be firing.

Then tested the wiring to the ignition coil number 2 and Found that the signal wire was not having signal to ignite the ignition coil. The signal wire was tested for continuity to the ECU and short circuits but was found to be OK. Hence the cause of no signal in the signal wire is the computer box [ECU] faulty.

WHAT CAUSED THE PROBLEM

- Faulty internal circuitry of the computer box [ECU]

THE FIX

- Replaced the computer box [ECU] and reprogrammed it

OPTIONAL FIX

- Repair the computer box [ECU]

(44) BANGING SOUND WHEN THE REVERSE OR DRIVE IS SELECTED

- Vehicle Make: VW Model:
- Golf V FSI Yea r: 2005-2008
- Engine code: BLX
- Steering: Right Hand Drive
- Transmission: Automatic

PROBL EM DESCRIPTION BY VEHICLE DRIVER

Vehicle was moving just fine then suddenly this problem appeared. Whenever I select the gear lever to reverse or drive there will be a very loud bang sound and the vehicle jerks.

SYMPTOMS

- Very loud bang sound from the gearbox when gear lever is moved to R (reverse) or D (drive)

HOW FAULT FINDING WAS DONE

- Diagnosed the car with a diagnostic scanner and retrieved the trouble codes:
  - P0715 input turbine or speed sensor A circuit malfunction
  - 00300 gearbox oil temperature sender (G93) short circuit to another valve
  - 00364 terminal 15 slave. no fault type identified
- Inspected the wire connectors on the gearbox. One connector was found to have two cables cut.
- Wiring diagram indicated that the two cables cut are for the input turbine speed sensor and the gearbox oil temperature sensor. Hence the cause of the problem

T HE FIX

- Joined the broken cables to their respective cut cables by soldering them

CHAPTER 3

ELECTRICAL PROBLEMS

(45) NO REVERSE LIGHT

- Vehicle Make: Ford
- Model: Fiesta
- Year: 2008-2011
- Engine Code: SPJA
- Steering Wheel: Right Hand Drive
- Transmission: Manual

PROB LEM DESCRIPTION BY VEHICLE DRIVER

The reverse light is not coming ON during reverse gear

SYMPTOMS

- No reverse light

HOW FAULT FINDING WAS DONE

- Located the reverse light switch on the gearbox
- Tested for 12-volt power to the reverse light switch connector (ok)
- Tested for 12 volt power on the reverse lamp while car is switched ON and someone engages reverse gear (no 12-volt power) -Tested for continuity in the wire between reverse light switch and reverse lamp (ok)
- Hence the problem is the reverse light switch

WHAT CAUSED THE FAULT

- Reverse light switch contacts found broken

THE FIX

- Replaced the reverse light switch with new one

(46) TEMPERATURE GAUGE NOT WORKING

- Vehicle Make: Toyota
- Model: Hilux

- Year: 1989-1993
- Eng i ne Code: 4 Y
- Steering Wheel: Right Hand Drive
- Transmission: Manual

PROBLEM DESCRIPTION BY VEHICLE DRIVER

The temperature gauge is not rising. It stays on minimum level even after driving for a long time.

SYMPTOMS

- Temperature gauge not rising
- Temperature gauge is on minimum constantly

HOW FAULT FINDING WAS DONE

- Located the temperature switch on the engine Dis connected the connector from the temperature switch. The disconnected wire
- bridged to the earth or the engine surface and immediately temperature gauge rose to the maximum. Hence the temperature switch condemned to be faulty

WHAT CAUSED THE FAULT

- Tear and wear of the temperature switch

TH E FIX

- Replaced the temperature switch with new one

(47) ONLY ONE HEADLIGHT FOR LOW BEAM IS WORKING

- Vehicle Make: Mazda
- Model: BT-50
- Year: 2006-2011
- Engine Code: MZR-CD
- Steering Wheel: Right Hand Drive
- Transmission: Manual

PROBLEM DESCRIPTION BY VEHICLE DRIVER

The left side low beam does not go ON even after replacing the lamp with new one. The low beam on the right side is working just fine.

SYMPTOMS

- Low beam on left side not switching ON
- Low beam on right side is working

HOW FAULT FINDING WAS DONE

- Tested for ground on the left side head lamp connector (OK)

- Tested for 12V positive power to low beam on the left side when low b eam switch is ON (Not OK. It reads 7V)

- Traced the low beam wire that was supposed to have 12V.The wire was continuous without any cut along it till it reached the fuse box, hence the cause of the 7V is the faulty internal parts of the fuse box. The fuse box is under the dash board on the driver's side. Both wires for the low beam headlights come from the fuse box. If the wire to the right headlight is ok with 12V but the one to the left headlight has 7V then the fuse box is faulty. The headlights switch for the low beam is ok since it's able to switch ON the right side low beam.

WHAT CAUSED THE FAULT

- Fuse box internal components malfunctioned. The fuse box is located on the fascia under the dashboard on the driver's side.

THE FIX

- Ins talled an independent circuit to supply the low beam for the left side. Included in the circuit is the relay and the 10A fuse

OPTIONAL FIX

- Replace the fuse box

(48) BRAKE LIGHTS STAY ON PERMANENTLY

- Vehicle Make: VW Model: Golf 5
- GTI Year: 2006 – 2008 Engine
- Code: BWA Steering Wheel: Right
- Hand Drive Transmission:
- Automatic
-

PROBLEM DESCRIPTION BY VEHICLE DRIVER

Noticed that the brake light stayed ON permanently, but they will go OFF when the car is OFF. When car is ON, all the 3 brake lights come ON permanently.

SYMPTOMS

- Bra ke lights stay ON permanently
- Brake light switch not responding

HOW FAULT FINDING WAS DONE

- Tested brake light switch (Ok)

- Disconnected the brake light wires from the multifunction control module (BCM) then the brake lights switched OFF. Hence the BCM internal circuit caused the problem

WHAT CAUSE THE FAULT

- BCM internal short circuit

THE FIX

- Rewired the brake light switch on its independent circuit with the fuse and relay bypassing the BCM.

OPTIONAL FIX

- Replace the BCM

(49) ALARM TURNS ON BY ITSELF

- Vehicle Make: Honda
- Model: CRV
- Year: 2002-2007
- Engine Code: K24A
- Steering Wheel: Right Hand Drive
- Transmission: Automatic

PROB LEM DESCRIPTION BY VEHICLE DRIVER

The alarm is always triggered a few minutes after locking the central locking system. This happens just by itself without anyone trying to open the doors.

SYMPTOMS

- All doors lock after locking them by remote
- Alarm is triggered by itself a few minutes after locking all doors

HOW FAULT FINDING WAS DONE

- Locked all the doors with remote control and immediately checked if all doors were locked. (OK)

- Waited for a few minutes after locking all the doors. After a few minutes alarm triggered itself without any one touching or getting closer to the vehicle. After the alarm triggered itself, only front passenger door was found unlocked while all other doors were still locked.

- Disc onnected the power supply from the front passenger door lock mechanism and locked all the doors again with remote control and waited for few minutes to see if alarm will be triggered. A few minutes passed but the alarm did not get triggered. Hence the front passenger door lock mechanism is faulty.

WHAT CAUSED THE FAULT

- Faulty front passenger door lock mechanism

THE FIX

- Replaced the front passenger door lock mechanism

(50) CE N TRAL LOCKING SYSTEM NOT WORKING

- Vehicle Make: Toyota
- Model: Hilux (2.5 D-4D)
- Year: 2005-2007
- Engine Code: 2KD
- Steering Wheel: Right Hand Drive
- Transmission: Manual

PROBLEM DESCRIPTION BY VEHICLE DRIVER

Only the driver's door is able to lock and unlock with the key manually. The other three doors cannot lock or unlock through central locking. Remote control is not available. Problem started after the a ftermarket alarm installation.

SYMPTOMS

- Only driver's door locks and unlocks with manual key
- The other three doors do not respond to central locking system

HOW FAULT FINDING WAS DONE

- Tested for 12 V power on driver's door lock motor while turning key i n the door key cylinder (ok)
- Tested for 12-volts on other three doors while turning key on the driver's door key cylinder. No 12 V power was present, hence the cause of the problem.

WHAT CAUSED THE FAULT

- Power supply to the three other door lock motors disconnected from driver's door lock motor power supply cable.

THE FIX

- Restored power supply to all the three door lock motors by connecting them in parallel with the power supply cable from driver’s door lock motor

(51) LOW COOLANT LEVEL PERMANENTLY SHOWING ON THE DASH BOARD

- Vehicle Make: VW
- Model: Golf IV
- Year: 1998 - 2004
- Engine Code: APK
- Steering Wheel: Right Hand Drive
- Tran smission: Automatic

PROBLEM DESCRIPTION BY VEHICLE DRIVER

The car was parked for a long time. Then after I started driving it again, th e low coolant warning light appeared permanently on the dash board even though the coolant level was full.

SYMPTOMS

- Low coolant warning light appearing permanently even though coolant is full
- Car drives fine without overheating

HOW FAULT FINDING WAS DONE

- Inspected the sensor contacts inside the coolant bottle. The sensor contacts were seen to be covered by dirty deposits, hence the cause of problem.

WHAT CAUSED THE FAULT

- Built up of dirty deposits around the sensor contacts

THE FIX

- Cleaned the sensor contacts

OPTIONAL FIX

- Replace the coolant bottle

(52) THE HORN SOUNDS VERY LOW

- Vehicle make: BMW
- Model: 3 series (F30)
- Year: 2012 - 2019
- Engine Code: B48 B20A
- Steering Wheel: Right Hand Drive
- Transmission: Automatic

PROBLEM DESCRIPTION BY VEHICLE DRIVER

The sound of the horn is very low. There is no sound coming out from the horn on the right side. Sound only comes out from the left horn.

SYMPTOMS

- No sound from the right horn

HOW FAULT FINDING WAS DONE

- Tested for positive 12 V power supply to the horn on the right as the horn switch is pressed. (Positive 12V is okay when horn switc h is pressed)

- Tested for ground/earth to the horn. Ground was found missing hence the cause of the problem

WHAT CAUSED THE FAULT

- Broken earth wire in the wiring harness to the horn on the right

THE FIX

- Tapped the earth/ground from nearby earth wires to supply the horn on the right

(53) FUEL GAUGE NOT WORKING

- Vehicle Make: Isuzu
- Mod el: KB 240
- Year: 2005 Engine
- Code: C24SE
- Steering Wheel: Right Hand Drive
- Transmission: Manual

PROB LEM DESCRIPTION BY VEHICLE DRIVER

The fuel gauge stopped working long time ago.

SYMPTOMS

- Fuel gauge not rising from minimum level

HOW FAULT FINDING WAS DONE

- Tested for continuity in the signal wire from fuel gauge sender unit to the instrument cluster. No continuity was found hence the cause of the problem

WHAT CAUSED THE FAULT

- Broken fuel guage signal wire in the wiring harness

THE FIX

- Sol dered new wire to bypass the section that was broken

(54) CAR SWITCHES OFF WHEN YOU INDICATE TO THE RIGHT

- Vehicle Make: VW
- Mo del: CADDY
- Year: 2005-2010
- Engine Code: BSF
- Steering Wheel: Right Hand Drive
- Transmission: Manual

PROB LEM DESCRIPTION BY VEHICLE DRIVER

While driving, it will switch off once I indicate to the right. This will only happen if I switch on TURN RIGHT indicator switch.

And then shortly after switching OFF it will then START again while it's in motion since it's a manual gearbox. When it goes OFF all lights on the dashboard will appear just like when car is only on ignition switch ON.

SYMPTOMS

- All systems are working ok and vehicle drives ok
- When indicator to the right is selected, car goes OFF and then comes back ON as you let go of the indicator switch.

HOW FAULT FINDING WAS DONE

- Removed the covers around the steering column to expose the steering column control module
- Turn key to ON then pushed the indicator switch to right and held it at that position, then immediately the dash board switched OFF. Then released the hand from the indicator switch then immediately the dashboard came ON again. Therefore as you push the indicator switch to the right the contacts between steering column control module and the steering column funct ion controls becomes loose causing the dash board to go OFF.

WHAT CAUSED THE FAULT

- Loose contacts between steering column control module and the steering column function controls

THE FIX

- Firmly secured the steering column control module so that contacts don't become loose again

OPTIONAL FIX

- Replace the steering column control module and the steering colum n function controls with new one

(55) BATTERY IS FLAT EVERY MORNING

- Vehicle Make: Mercedes
- Model: C-CLASS (203)
- Year: 2000-2006
- Eng ine Code: 112. 946
- Steering Wheel: Right Hand Drive
- Transmission: Automatic

PROB LEM DESCRIPTION BY VEHICLE DRIVER

Every morning the battery is flat. Car needs to be jump started only in the morning then the rest of the day it will be ok. But the following morning it must be jump started.

SYMPTOMS

- Dashboard lights come ON as normal
- Cranking is very weak

HOW FAULT FINDING WAS DONE

- Tested for parasitic battery drain. The current drawn during parasitic battery drain test was 0.85 amps. The 0.85 amps parasitic battery drain causes the battery to drain overnight. Rem oved fuse F27 then current draw dropped to 0.03 amps. 0.03
- amps is normal. Fuse F27 supplies constant power to the front passenger seat control module.

WHAT CAUSED THE FAULT

- Front passenger seat control module internal circuitry faulty

THE FIX

- Rewired the constant power supply to the seat control module incorporating a relay to be switched OFF once the car is OFF and ON when the car is switched ON.

OPTIONAL FIX

- Replace the front passenger seat control module

  OR

- Remove fuse F27

(56) RADIATOR FAN RUN CONTINUOUSLY

- Vehicle: Make Mazda
- Model: Mazda 3 (Axela)
- Year: 2003-2009
- Engine Code: Z6
- Steering Wheel: Right Hand Drive
- Transmission: Automatic

PROBLEM DESCRIPTION BY VEHICLE DRIVER

The radiator fan start running continuously once the car is switched ON. The temperature gauge stays permanently on low temperature.

SYMPTOMS

- Radiator fan running non-stop
- Temperature gauge not rising

HOW FAULT FINDING WAS DONE

- Inspected wiring to engine coolant temperature sensor and found that connector was broken and plugged out of the sensor. Hence the cause of the problem

WHAT CAUSED THE FAULT

- Broken connector to ECT (engine coolant temperature sensor)

THE FIX

- Replaced the broken connector

(57) STEERING WHEEL HARD TO TURN

- Vehicle Make: VW
- Model: Polo (cross polo)
- Year: 2007-2009
- Engine Code: BTS
- Steerin g Wheel: Right Hand Drive
- Transmission: Automatic

PROBLEM DESCRIPTION BY VEHICLE DRIVER

Vehicle got involved in an accident which led to damage of the left Fender. After accident and after Panel Beating, power steering was found to be hard to turn.

SYMPTOMS

- Steer ing wheel hard to turn
- Steering wheel warning light ON

HOW FAULT FINDING WAS DONE

- Tested for 12 volt constant power to power steering control module (ok)
- Tested for ground to power steering control module (ok)
- Tested for ignition switch ON power supply to steering control module (ok)
- Tested for communication wires to power steering control module (ok)
- All tests found to be ok, hence the power steering unit is the cause of the problem
- Scanned the steering assistance control module with a diagnostic scanner and retrieved trouble code: 01656 – crash signal.
- (The crash signal trouble code stored is the one causing the problem)

WHAT CAUSED THE FAULT

- Crash signal trouble code stored

THE FIX

- Cleared the trouble code (01656-crash signal) with diagnostic equipment

58) SWITCHES OFF DURING DRIVING

- Vehicle Make: Volvo
- Model: V40
- Year: 2013

- Engine Code: B4164
- Steeṅng Wheel: Right Hand Drive
- Transmission: Automatic

PROBLEM DESCRIPTION BY VEHICLE DRIVER

I bought the car yesterday as good second hand from second hand car dealer. While driving I noticed the battery warning light appearing on the dashboard. Then shortly the steering wheel became hard to turn. Then shortly the car just switched OFF.

SYMPTOMS

- Battery light appearing while driving
- Steering wheel becomes hard to turn
- Car switches OFF while driving and refuses to START

HOW FAULT FINDING WAS DONE

- Inspected the battery and observed that the battery was leaking acid. Battery leaking acid is an indication that the battery is faulty/dead.
- Replaced the leaking battery with another battery and STAR TED the car. While car is idling I tested for overcharging, undercharging and normal battery charging. The tests proved that the charging system was working just fine. Hence the problem was the battery gone bad.

WHAT CAUSED THE FAULT

- Battery dead due to age

THE FIX

- Replaced the battery with the new one that has the correct amp hour(AH) rating.Amp Hour of the battery of this car is 85AH.

(59) TAIL LIGHTS NOT WORKING

- Vehicle Make: Toyota
- Model: VIOS
- Year: 2001-2006
- Engine Code: 1NZ
- Steering Wheel: Right Hand Drive
- Transmission: Manual

PROBLEM DESCRIPTION BY VEHICLE DRIVER

The tai l lights, both of them are not working

SYMPTOMS

- Brake lights are working
- Tail lights not working

HOW FAULT FINDING WAS DONE

- Removed the tail lamps and inspected both of them. The tail lamps found burnt on their filaments

WHAT CAUSED THE FAULT

- Tail lamps lifespan exceeded

THE FIX

- Replaced the tail lamps

(60) FUEL GAUGE IS NOT WORKING

- Vehicle Make: FORD Model:
- FOCUS Year: 1998-2005 Engine
- Code: FYDA Steering Wheel:
- Right Hand Drive Transmission:
- Automatic
-

PROBLEM DESCRIPTION BY VEHICLE DRIVER

The fuel gauge stopped working suddenly

SYMPTOMS

- The fuel gauge only ends below quarter level

HOW FAULT FINDING WAS DONE

- Removed fuel pump housing and inspected fuel gauge sender unit. The sender unit was found deteriorated on its variable resistor

WHAT CAUSED THE FAULT

- Tear and wear of the variable resistor on the fuel guage sender unit

THE FIX

- Replaced the complete fuel pump housing with new one

(61) LEFT TAIL LAMP SWITCHES OFF WHEN BRAKE IS DEPRESSED

- Vehicle Make: Toyota
- Model: VIOS
- Year: 2001-2006
- Engine Code: 1NZ
- Steering Wheel: Right Hand Drive
- Transmission: Manual

PROBLEM DESCRIPTION BY VEHICLE DRIVER

The tail lamp on the left goes off when I brake but the right lamp will still be on

SYMPTOMS

- Left tail lamp goes OFF during braking
- The right tail lamp still remains ON while left tail lamp goes OFF during braking

HOW FAULT FINDING WAS DONE

- Remo ved the tail lamp fitting and inspected it. The Point of connection of Earth wire to lamp fitting was found rusted, hence the cause of problem.

WHAT CAUSED THE FAULT

- Rust build-up around lamp fitting

THE FIX

- Replaced the lamp fitting

(62) BRAKE LIGHTS NOT WORKING

- Vehicle Make: Kia
- Model: Cerato
- Year: 2019-2021
- Engine Code: G4FG
- Ste ering Wheel: Right Hand Drive
- Transmission: Manual

PROBLEM DESCRIPTION BY VEHICLE DRIVER

Brake lights were suddenly noticed that they stopped working

SYMPTOMS

- Both brake lights are not working
- Both tail lights are working

HOW FAULT FINDING WAS DONE

- Inspected the brake lamps both were found burnt

WHAT CAUSED THE FAULT

- Lamps worn out due to tear and wear

THE FIX

- Replaced both lamps with new ones

(63) FRONT DRIVER'S SEAT ADJUSTMENTS NOT WORKING

- Vehicle Make: Audi
- Model: A4 (Quattro) (2TFSI)
- Year: 2006-2008
- Engine Code: BWE
- Steering Wheel: Right Hand Drive
- Transmission: Automatic

PROBLEM DESCRIPTION BY VEHICLE DRIVER

The vehicle was recently purchased with this problem existing from the second-hand car dealer

SYMPTOMS

- All the adjustments on the driver's seat are not working

HOW FAULT FINDING WAS DONE

- Tested for the ground to the main seat adjustment control module on the side of the seat (ok) Tested 12 volt power supply
- to the main seat adjustment control module on the side of the seat, 12 volts power supply missing, hence the cause of the problem

WHAT CAUSED THE FAULT

- Fus e blown due to sudden over current drawn

THE FIX

- Restored power supply to the control module

(64) STEERING WHEEL HARD TO TURN

- Vehicle Make: Toyota
- Model: Run X
- Year: 2001-2006
- Engine Code: 1NZ
- Steering Wheel: Right Hand Drive
- Transmission: Automatic

PROB L EM DESCRIPTION BY VEHICLE DRIVER

I went to some mechanic to have the radiator fan fixed since it had stopped running. The mechanic connected the radiator fan direct to one of the power supply wires in the wiring harness. I then noticed that the steering wheel became hard to turn after the mechanic finished connecting the radiator fan. Before I took the car to the mechanic, the steering wheel was just fine.

SYMPTOMS

- P/S warning light showing on the dash board
- Steering wheel hard to turn

HOW FAULT FINDING WAS DONE

- Tested the power supplies to the steering control module (OK)
- Tested the ground to the steering control module (OK) Tested
- the power steering motor (OK) Tested the power that was being
- supplied to the motor by the steerin g control module while turning the steering wheel to the right and left as the car was idling. The tested power was found to be fractuating over 19 volts.Hence the power steering control module is faulty since the power steering motor is a 12 volts rated motor.

WHAT CAUSED THE FAULT

- Faulty internal circuitry of the steering control module

THE FIX

- Replaced the steering control module

(65) WHEN THE CAR IS HOT IT WILL NOT START

- Vehicle Make: VW
- Model: Polo (9N1/9N2/9N3)
- Year: 2003-2006
- Engine Code: BBY
- Steering Wheel: Right Hand Drive
- Transmission: Automatic

PROBLEM DESCRIPTION BY VEHICLE DRIVER

When the car is cold and the weather outside is also cold, the car is able to start. After driving the car for a long time and it becomes hot, when you switch off and try to start the car it will not start but starts blinking the key warning symbol. When key symbol starts blinking it will only crank but will not start. Sometimes it will not start even when the car is cold but as long as the outside air is hot from the sun the key will blink making it not to start the car.

SYMPTOMS

- When engine is hot key symbol will be blinking when trying to start the car.
- The car will not start when outside its sunny and hot, key symbol will be blinking when trying to start the car

- All the lights on the dashboard switches ON as normal
- Whe n the key symbol start blinking, car will only try to start but goes OFF immediately

HOW FAULT FINDING WAS DONE

- Opened the instrument cluster and inspected the soldering on its motherboard. Some soldered points on the motherboard where found slightly cracked, hence the cause of the key blinking

WHAT CAUSED THE FAULT

- Cracked soldered points on the motherboard of the instrument cluster

THE FIX

- Resoldered the solder points which had cracks around them on the motherboard

(66) RADIATOR FANS NOT RUNNING

- Vehicle Make: Audi
- Mod el: A4
- Year: 2001-2008
- Engine Number: BDV
- Steering Wheel: Right Hand Drive
- Transmission: Automatic

PROB LEM DESCRIPTION BY VEHICLE DRIVER

The radiator fan does not run. Even when AC is switched ON still it will not run.

SYMPTOMS

- Temperature gauge rising beyond normal
- Coolant in expansion water bottle boils

HOW FAULT FINDING WAS DONE

- Unplugged engine coolant sensor and started the car. The radi ator fan did not run. (It is supposed to run when engine coolant temp (ECT) sensor is unplugged)

- Plugged ECT back and started the car. Then switched ON the AC but radiator fan did not run. (Supposed to run when AC is on)

- Tested for 12 volt positive power supply to fan control module (OK)

- Tested for negative/ground to the fan control module (OK)

- Tested both the fan motors by bypassing the fan control module (OK) hence the fault is on the fan control module

WHAT CAUSED THE FAULT

- Faulty internal circuitry of fan control module

THE FIX

- Replaced the fan control module

(67) RADIATOR FAN RUNS CONTINUOUSLY

- Vehicle Make: Mazda
- Model: axela
- Year: 2009-2014
- Eng ine Code: ZY
- Steering Wheel: Right Hand Drive
- Transmission: Automatic

PROBLEM DESCRIPTION BY VEHICLE DRIVER

Radiator fan runs continuously without stopping. It will only stop running when any battery terminal is removed from the battery.

SYMPTOMS

- Radiator fan runs continuously
- Radiator fan is running continuously at its highest speed

HOW FAULT FINDING WAS DONE

- Scanned the computer box (ECU) with diagnostic equipment but no trouble code was stored in the computer box.
- Disconnected the signal wire to the fan control module but still the fan would run continuously hence the fan control module condemned

WHAT CAUSED THE FAULT

- Faulty internal circuitry of the fan control module

THE FIX

- Repl aced the fan control module together with the fan since it's one assembly

OPTIONAL FIX

- Repla ce only the fan control module

(68) NOT CHARGING THE BATTERY

- Vehi cle Make: VW
- Model: Polo (6R1)
- Year: 2009-2014
- Engine Code: CGG
- Steering Wheel: Right Hand Drive

- Transmission: Automatic

PROBLEM DESCRIPTION BY VEHICLE DRIVER

While driving, the battery light appeared on the dashboard permanently

SYMPTOMS

- Battery charge indicator lamp comes ON and does not go OFF after starting the car.
- After driving for a long distance, the battery runs flat and the car switches OFF.

HOW FAULT FINDING WAS DONE

- Tested for 12 volts positive constant to the alternator output pin (ok)
- Tested for some power on the two pins (blue and brown wires) that connect to the voltage regulator.
- There is some amount of power in the blue wire on the connector. The brown wire on the connector also has some amount of power. The voltage found in both blue and brown wires is not 12 volts it is between 4 volts and 10 volts. The voltage in the blue and brown wires on the voltage regulator

connector appears after ignition is switched ON. Hence the problem is on the alternator as wires prove to have some voltage in them.

WHAT CAUSED THE FAULT

- Tear and wear of the alternator

THE FIX

- Replaced the alternator with the working one

OPTIONAL FIX

- Test the alternator on the bench and repair it.

(69) HORN NOT WORKING

- Vehicle Make: Audi
- Model: A3 (FSI)
- Year: 2003-2008
- Engine Code: AXW
- Steering Wheel: Right Hand Drive
- Transmission: Automatic

PROBLEM DESCRIPTION BY VEHICLE DRIVER

The horn suddenly stopped working though yesterday it was working quite ok

SYMPTOMS

- No sound of horn relay clicking when pressing the steering wheel airbag

HOW FAULT FINDING WAS DONE

- Removed covers around the steering column. Steering column control module found loosely plugged on one of its connectors, hence the cause of the problem

WHAT CAUSED THE FAULT

- Steering column control module connectors not firmly plugged into place due to tear and wear

THE FIX

- Firmly plugged the loose connector of the steering column module onto its corresponding connector

OP TIONAL FIX

- Replace steering control module with one that has good clipping connectors

(70) NO BRAKE LIGHTS

- Vehicle Make: Audi
- Model: A4
- Year: 2001-2008
- Engine Code: BDV
- Steering wheel: Right Hand Drive
- Transmission: Automatic

PROBLEM DESCRIPTION BY VEHICLE DRIVER

Brake lights no longer work after car was parked for about two years

SYMPTOMS

- Brake lights not lighting when brake pedal is depressed

HOW FAULT FINDING WAS DONE

- Tested the brake switch for continuity in two of its four pins when operated (not Ok), hence the cause of the problem

WHAT CAUSED THE FAULT

- Malfunction brake light switch

THE FIX

- Replaced the brake light switch with the new one

(71) NOT STARTING BUT ONLY CRANKS

- Vehicle Make: VW
- Model: Polo (TDI)
- Year: 2009-2011
- Engine Code: CLN
- Steering Wheel: Right Hand Drive
- Tran smission: MANUAL

PROBLEM DESCRIPTION BY VEHICLE DRIVER

While dri ving the car, it slowly decelerated itself and at the same time the glow plug warning light symbol started flashing on the dashboard. Deceleration continued till the car switched OFF by itself.

SYMPTOMS

- Dashboard lights come ON as normal
- Glow plug warning light does not stop flashing

HOW FAULT FINDING WAS DONE

- Scanned the vehicle engine management system. The trouble code retrieved was: P0684 GPCM to PCM communication circuit range/performance Checked the fuse for glow plug
- control module on the fuse box under the dashboard. Glow plug control module fuse F31 missing on its place. Hence the cause of problem

WHAT CAUSED THE FAULT

- Fuse F 31 missing (it is assumed that the fuse was loose on its place and it fell off while driving)

THE FIX

- Insta ll fuse on F31

(72) STEERING WHEEL HARD TO TURN

- Veh icle Make: VW
- Model: Polo (9N1/9N2/9N3)
- Year: 2002-2006
- Engine Code: BBY
- Steering Wheel: Right Hand Drive
- Tra nsmission: Automatic

PROBLEM DESCRIPTION BY VEHICLE DRIVER

The car was recently bought with steering wheel hard to turn

SYMPTOMS

- . Steering wheel hard to turn .
- Steering wheel symbol appearing on dashboard

HOW FAULT FINDING WAS DONE

- Tested for constant power supply to the steering pump module (ok)
- Tested for ignition switch ON power supply to the steering pump module (ok)
- Tested for earth to the steering pump module (ok)
- All tests proved to be ok, hence the power steering module is condemned
- Steering pump is located in front of the left front wheel after removing the mud guard

WHAT CAUSED THE FAULT

- Tear and wear of the steering pump control module/the steering pump motor

THE FIX

- Replaced the power steering pump complete unit

OPTIONAL FIX

- Repair the steering control module/the steering pump motor according to what is found to be faulty

(73) ETCS FUSE KEEPS ON BLOWING UP

- Vehicle Make: Toyota
- Model: Yaris
- Year: 2001-2006
- Engine Code: 1NZ
- Steering Wheel: Right Hand Drive
- Transmission: Automatic

PROBLEM DESCRIPTION BY VEHICLE DRIVER

The fuse for the ETCS keeps on blowing up even when the car is OFF

SYMPTOMS

- ETC S blowing up

HOW FAULT FINDING WAS DONE

- Removed the blown up ETCS fuse and replaced it with a good one but immediately it blew up
- Disconnected battery negative then unplugged the engine control module from its connectors. Replaced blown up ETCS fuse with a good one then plugged the engine control module back on its connectors. Reconnected the battery negative and observed if the ETCS fuse will blow up but it did not blow up. Hence the cause of the ETCS blowing up is the engine control module.

- Disconnected battery negative and unplugged the engine con t rol module to inspect its circuit board. One diode on the circuit board was found burnt hence causing the short circuit for ETCS to blow up.

WHAT CAUSED THE FAULT

- Burnt diode on the engine control module circuit board

THE FIX

- Repaired the engine control module

OPTINAL FIX

- Replace the engine control module

(74) RADIATOR FAN RUNS CONTINUOUSLY

- Vehicle Make: Audi
- Model: A3 (FSI)
- Yea r: 2005-2008
- Engine Code: BLR
- Steering Wheel: Right Hand Drive
- Transmission: Automatic

PROBLEM DESCRIPTION BY VEHICLE DRIVER

Radiator fan runs continuously without stopping even when you turn the ignition switch OFF. To stop the fan you need to remove one of the battery terminals. This problem just started from nowhere.

SYMPTOMS

- Both the big and small radiator fan run continuously without stopping
- Battery terminal must be removed to stop the fans running

HOW FAULT FINDING WAS DONE

- Tested the power supply in the big wire to the fan control module (ok)
- Tested for short circuits on the small wires to fan control module (no short circuits)
- Tested for ground to the fan control module (ok)
- All the tests proved to be OK hence the fan control module is condemned

WHAT CAUSED THE FAULT

- Internal short circuits in the fan control module

THE FIX

- Replaced the big radiator fan as one unit with fan control module

(75) HEADLIGHT DIM (LOW BEAM) NOT SWITCHING OFF

- Vehicle Make: VW
- Model: Jetta
- Year: 2005-2011
- Engine Code: BLR (FSI)
- Steering Wheel: Right Hand Drive
- Transmission: Automatic

PROBLEM DESCRIPTION BY VEHICLE DRIVER

The dim lights stay ON continuously without switching OFF. The bright (high beam) do not switch ON at all. The first thing that stopped working was the wipers. I noticed some kind of sparks coming out from the wipers motor on the front wind shield. The sparks stopped and I continued driving, then when I parked I realized that the low beam lights were ON even when their switch was OFF.

SYMPTOMS

- Dim (low beam) headlights don't switch OFF
- Bright (high beam) headlights don't switch ON
- Wipers not working

HOW FAULT FINDING WAS DONE

- Opened the multifunction control module/bcm (body control module)
- One component on the bcm printed circuit board was found burnt

- Thebcm is located under the steering wheel

WHAT CAUSED THE FAULT

- Sparks coming out from the wiper motor (burning of wiper motor)

THE FIX

- Replaced the BCM and the wiper motor

(76) NO HORN

- Veh icle Make: Mercedes Benz
- Model: C- CLASS (204) C200 kompressor
- Year: 2007-2010
- Engine code: 271.950
- Steering Wheel: Right Hand Drive
- Tra nsmission: Automatic\

PROBLEM DESCRIPTION BY VEHICLE DRIVER

Before th e steering lock was removed for repairs by some mechanic the horn was working just fine. When the steering lock came back and was installed back in its place that's when the horn stopped working.

SYMPTOMS

- No sound of horn when steering wheel air bag is pressed
- All switches on the steering column are working just fine

HOW FAULT FINDING WAS DONE

- Battery negative disconnected and removed airbag to access the clock spring. Removed the clock spring and inspected the ribbon
- inside. The ribbon was found broken on one of its wires hence the cause of no horn.

WHAT CAUSED THE FAULT

- Previous mechanic broke the ribbon inside the clock spring

THE FIX

- Repaired the clock spring

OPTIONAL FIX

- Replace the entire steering column switches assembly. Clock spring is on the steering column switch assembly.

(77) NOT CRANKING

- Vehicle Make: Toyota
- Model: CAMRY
- Year: 2001-2006
- Engine Code: 2AZ-FE
- Steering Wheel: Right Hand Drive
- Transmission: Automatic

PROBLEM DESCRIPTION BY VEHICLE DRIVER

The car was parked yesterday and it was just fine. The following day when I tried to START, it did not crank.

SYMPTOMS

- Not cranking
- All dashboard lights switch ON as normal

HOW FAULT FINDING WAS DONE

- Tested for constant 12 volts power to starter solenoid (ok)
- Disconnected small cable from starter solenoid
- Test ed for 12volts power in the small cable that is disconnected as someone is trying to start the car. (Ok)
- Hence the problem is on the starter or solenoid

WHAT CAUSED THE FAULT

- Tear and wear of the starter or solenoid

THE FIX

- Replaced the whole starter unit

OPTIONAL FIX

- Repair the starter unit

(78) RADIATOR FAN NOT RUNNING

- Vehicle Make: VW
- Model: Polo (9N1/9N2/9N3)
- Year: 2005-2009
- Engine Code: BKY
- Ste ering Wheel: Right Hand Drive
- Transmission: Automatic

PROBLEM DESCRIPTION BY VEHICLE DRIVER

Suddenly the radiator fan just stopped working. Water in the bottle started boiling. Temperature gauge rising close to approaching maximum level.

SYMPTOMS

- Temperature gauge reading going beyond normal
- Water in the water bottle boiling
- Radiator fan not running

HOW FAULT FINDING WAS DONE

- Started the car and tested if the thermostat was working (ok) Inspected for any water leaks (no leaks) Unplug the
- connector on the radiator thermal switch. Then connected a jumper wire from the battery negative and started placing it onto the three pins on the connector, one pin at a time. When wire was placed onto the correct pin, the radiator fan s tarted running. Hence the cause of the problem is the thermal switch on the radiator.

WHAT CAUSED THE FAULT

Tear and wear of the radiator thermal switch

THE FIX

- Replaced the radiator thermal switch with the new one

(79) HAZARD AND CENTRE CONSOLE FUNCTIONS NOT WORKING

- Vehicle Make: Toyota
- Model: Run X
- Year: 2001-2006
- Engine Code: 1NZ
- Steering Wheel: Right Hand Drive
- Transmission: Automatic

PROBLEM DESCRIPTION BY VEHICLE DRIVER

I just noticed that the hazard and all the functions on the centre console stopped working.

SYMPTOMS

- All hazard lights not working but individual indicators working
- All c entre console functions not working I,e clock, A/C etc.

HOW FAULT FINDING WAS DONE

- Removed the centre console to access the power supply conne ctor to the center console. Two connectors to the centre console were noticed to be unplugged out from the center console. Hence the cause of the problem

WHAT CAUSED THE FAULT

- Connectors unplugged out from the centre console.

THE FIX

- Firmly replugged back the connector to the centre console

(80) ALTERNATOR NOT CHARGING

- Vehicle Make: Audi
- Model: A3 (FSI)
- Year: 2005-2008
- Engine Code: BLR
- Steering Wheel: Right Hand Drive
- Transmission: Automatic

PROBLEM DESCRIPTION BY VEHICLE DRIVER

Alternator is not chargingthe battery. Car willswitch OFF after driving for some distancedue to battery gone flat.

SYMPTOMS

- Battery light showing on the dashboard
- Battery goes flat while driving car
- Car goes off if one battery terminal is disconnected from the

battery while car is idling

HOW FAULT FINDING WAS DONE

- Measured voltage across battery terminals with multimeter while car is idling. Less than 12V shows on the multimeter. Voltage is supposed to be 13.5 to 14.5 volts if charging system is working. Tested for constant 12V power supply to the alternator. (No
- Power) While car is off and battery negative removed, tested for continuity on the 200 amp alternator fuse F1. No continuity
- across the fuse. Fuse was found blown hence the cause of the problem.

WHAT CAUSED THE FAULT

- Overcurrent drawn through the fuse

THE FIX

- Repl aced the blown fuse with the fuse of the same amp rating
- Alternator fuse is on the fuse box next to the battery

(81) NOT STARTING

- Vehicle Make: VW
- Mod el: Golf V (FSI)
- Year: 2004
- Engine Code: AXW
- Steering Wheel: Right Hand Drive
- Transmission: Automatic

PROBLEM DESCRIPTION BY VEHICLE DRIVER

The way the problem started, when you try to start it would be like it wants to start but goes OFF. Till it reached a point that it only cranks and would not even try to start.

SYMPTOMS

- Dashboard light switches ON as normal
- Cranks but no start
- Fuel pump can be heard running when ignition switch is turned to ON

HOW FAULT FINDING WAS DONE

- Disconnected the outlet pipe from the fuel pump Conn ected the
- fuel pressure gauge on the outlet of the fuel pump Asked someone to crank the car while I observed the pressure readings
- on the fuel pressure gauge. Less than 1 bar was indicated on the gauge as the car cranked. Fuel pressure on this

  car is supposed to be 4.8 bar as fuel delivery pressure to the mechanical high pressure pump on the engine. Hence fuel pump is condemned.

WHAT CAUSED THE FAULT

- Fuel pump tear and wear

THE FIX

- Replaced the fuel pump with new one complete with the housing

(82) NOT CHARGING

- Vehicle Make: VW
- Model: GOLF V (FSI)
- Year: 2004
- Engine Code: AXW
- Steering Wheel: RightHandDrive
- Transmission: Automatic

The vehicle will switch OFF due to battery gone flat after driving for some distance.

SYMPTOMS

- Car will switch OFF after driving for a while
- Battery light shows on the dashboard

HOW FAULT FINDING WAS DONE

- Measured voltage across battery terminals with multimeter while car was idling. Less than 12V showed on the multimeter. Voltage is supposed to be 13.5 to 14.5 volts if charging system is working. Tested for constant 12V power supply to the
- alternator. (No Power)
- While car is off and battery negative removed, tested for continuity on the 200 amp alternator fuse F1. No continuity across the fuse. Fuse is blown, hence the cause of the problem.

WHAT CAUSED THE FAULT

- Overcurrent drawn through the fuse

THE FIX

- Replaced the blown fuse with the fuse of the same amp rating
- Alternator fuse is on the fuse box next to the battery

(83) NOT STARTING

- Vehicle Make: MERCEDEZ BENZ
- Model: C-CLASS (204) C180 Kompressor
- Year: 2007-2009
- Engine Code: 271.952
- Stee ring Wheel: Right Hand Drive
- Transmission: Automatic

PROBLEM DESCRIPTION BY VEHICLE DRIVER

The car was driving just fine then parked it. For about two weeks I did not use the car so when I tried to start it, the car refused to start. The first time it stopped starting, when I open the door, the mileage would show on the dashboard. But after the other mechanic tried to fix the car, even the mileage has now stopped showing on the dashboard when you open the door.

SYMPTOMS

- No sound from steering lock as key fob is inserted into the ignition switch

- Key is able to turn but dashboard does not switch ON
- No c ranking when key is turned to START

HOW FAULT FINDING WAS DONE

- Battery negative removed then removed all connectors from the fuse box (Fuse box in the engine bay on the right) With battery reconnected, tested for 12V constant power on connector with pin 12S (no power) Battery negative removed then accessed the
- fuses on power distribution box next to the battery to test the fuses for continuity. Fuse FL 9 has no continuity indicating that fuse is blown. Replaced fuse FL 9 and reconnected the battery. Then opened the driver's door, dashboard illuminated and
- mileage appeared. Insert ed the key fob and started the car, car started but switched OFF immediately. Tried second time, it started but switched OFF immediately. Then third time refused
- to start completely. Tested for power supply and grounds to steering lock module

-

and e lectronic ignition switch (ok)
- Tested for any voltage fluctuations on the communication wires on the steering lock and ignition switch as key fob is inserted into the ignition (no voltage fluctuations, hence key fob malfunctioned)

WHAT CAUSED THE FAULT

- Initial problem was key fob malfunctioned but the previous mechanic caused the second problem of blowing up fuse FL9

THE FIX

- Replaced fuse FL 9
- Reprogrammed the key fob

(84) BATTERY BEING DRAINED

- Vehicle Make: MERCEDEZ BENZ
- Model: C-CLASS, (203) C240
- Year: 2000-2006
- Engine Code: 112.912
- Steering Wheel: Right Hand Drive
- Transmission: Automatic

PROBLEM DESCRIPTION BY VEHICLE DRIVER

The car c ranks very weak in the morning such that I will need to jump start it. This happens if I start the car late morning after 10 am but if I start the car very early e.g. at 8 am the car will start. Once the car starts, I can drive the car throughout the day with no problem.

SYMPTOMS

- Cranks very weak in the morning
- Car needs to be jump started in the morning

HOW FAULT FINDING WAS DONE

- Removed battery negative and connected multimeter in series with the battery to test for parasitic battery drain while car is OFF. Multimeter set at 10 amp reading and probes connected the correct way

- All three doors closed but driver's door open and in closed posi t ion on its door lock mechanism. While multimeter in
- series with battery, then reconnected the battery negative. Current reading on the multimeter was 1.20 amp then dropped to 1.03amp and stayed permanently at 1.03amp.1.03 amp is too much current draw when car is OFF hence the cause of battery drain. Removed fuse F42
- then the current draw dropped to 0.09.F42 is on the fuse box on the fascia under the far right of the steering wheel.F42 is supplying constant power to dashboard (instrument cluster).Disconnected the power connectors from the instrument cluster and placed back F42 on its place but current draw stayed at 0.09, hence the instrument cluster is draining the battery through constant power.
- Malfunctioned internal circuitry of the instrument cluster causing battery drain

THE FIX

- Reconnected 12V constant power to instrument cluster via an independent relay. The relay activates the constant power when ignitionswitchis turned to ON.

OP TIONAL FIX

- Replace instrument cluster with new one.

(85) THE DIM (LOW BAEM) HEADLIGHTS NOT GOING OFF

- Vehicle Make: Toyota
- Model: COROLLA

- Year: 2000 - 2007
- Eng i ne Code: 1NZ
- Steering Wheel: Right Hand Drive
- Transmission: Automatic

PROBLEM DESCRIPTION BY VEHICLE DRIVER

The headlights for the low beam (dim lights) stays ON permanently without switching OFF even when the low beam switch is switched OFF. To switch them OFF I need to always remove their fuse.

SYMPTOMS

- Low beam headlights (dim lights) not switching OFF
- All other lights working OK as normal
- Dim lights remain ON even when the car is switched OFF

HOW FAULT FINDING WAS DONE

- Removed the power supply connector from the dim (low beam) headl ights, right and left respectively Tested for 12V constant
- power supply on the connectors for head light (ok). The lamp is the three pin H4 lamp which is supplied by constant 12V power
- supply on one pin only. The other two pins

  receiv e grounds from the combination switch for dim and bright.

  Removed the connector from the combination switch
- Tested for ground/earth on the pin for bright (no ground)
-

- Tested for ground/earth on the pin for dim (ground is prese n t).Ground is not supposed to be present since combination switch is disconnected, that means the ground wire to the dim lights is short circuited to ground somewhere in the wiring harness hence the cause of the problem.

WHAT CAUSED THE FAULT

- Ground wire insulation damaged in the wiring harness

THE FIX

- Rewired a separate wire from the combination switch to both low beam lights bypassing the short circuited wire

CHAPTER 4

SUSPENSION PROBLEMS

(86) NOISE OR SOUND DURING TURNING TO THE RIGHT OR LEFT

- Vehicle make: VW
- Model: Golf V FSI
- Year: 2005-2008
- Engine code: BLX
- Transmission: Automatic
- Steering: Right Hand Drive

SYMPTOMS

- Noise from the left wheel during turning to the right or left

HOW FAULT FINDING WAS DONE

- Inspected the CV joints on the left front wheel. The outer CV joint was found to be damaged

WHAT CAUSED THE FAULT

- Tear and wear of the outer CV joint

TH E FIX

- Replaced the left outer CV

OPTIONAL FIX

- Replace all the inner and outer cv joints of both wheels

(87) REAR WHEELS SWEEPING THE CAR OUT OF THE ROAD

- Vehicle Make: VW
- Model: Polo
- Year: 2005-2009
- Engine code: BKY
- Transmission: Automatic
- Steering: Right Hand Drive

PROBLEM DESCRIPTION BY VEHICLE DRIVER

One nigh t when I was driving fast, I hit the kerbstone with the left rear wheel. That is when the problem started. The rear wheels are sweeping the rear of the car towards the left, sliding the car out of the road.

SYMPTOMS

- Rear wheels sweeping the car to the left during driving

H OW FAULT FINDING WAS DONE

- Inspected the left rear wheel and found it out of alignment. The wheel is protruding out of the mudguard

WHAT CAUSED THE FAULT

- Rear left wheel out of alignment

THE FIX

- Replaced the entire rear axle

(8 8) HOWLING SOUND FROM THE REAR RIGHT WHEEL

- Vehicle Maker: VW
- Model: Golf V
- Year : 2006-2008
- Engine code: BWA
- Transmission: Automatic
- Steering: Right Hand Drive

PROB LEM DESCRIPTION BY VEHICLE DRIVER

I suddenly started hearing some howling sound coming from the right rear wheel.

SYMPTOMS

- Howling sound coming from the wheel during driving

HOW FAULT FINDING WAS DONE

- Inspected the right rear wheel hub. The bearing inside the hub was seen to be damaged. Hence the cause of the problem

THE FIX

- Replaced the entire wheel hub with the new one.

(89) HOWLING SOUND FROM THE LEFT FRONT WHEEL

- Vehicle Make: VW
- Model: Golf IV
- Year: 2000-2003
- Eng ine code: AVU
- Steering: Right Hand Drive
- Transmission: Automatic

PROBLEM DESCRIPTION BY VEHICLE DRIVER

The car was parked for a very long time close to two years due to some engine problem. After the car was fixed, during test driving I could hear some howling sound coming from the front left wheel.

SYMPTOMS

- When you increase the speed, front left wheel feels like it is continuously passing through small many humps or it feels like its wobbling

HOW FAULT FINDING WAS DONE

- Removed the tire and inspected the wheel bearing. The wheel bearing was felt to be very loose when wiggling or shaking it.
- After removing the wheel hub and inspected the wheel bearing, grease coming out from the wheel bearing could be seen. Hence the cause of the problem.

WHAT CAUSED THE FAULT

- Tear and wear of the wheel bearing

THE FIX

- Replaced the complete wheel hub with a good second hand.

OPT IONAL FIX

- Replace the wheel bearing only with new one

(90) HOWLING SOUND FROM THE REAR LEFT WHEEL

- Vehicle Make: VW
- Model: Golf V, GTI
- Year: 2006-2008
- Engine code: BWA

- Steering wheel: Right Hand Drive
- Tran smission: Automatic

PROBLEM DESCRIPTION BY VEHICLE DRIVER

There's a howling noise coming from the left rear wheel. The noise increases with increase in speed. The noise started at low pitch but after some months it became so loud.

SYMPTOMS

- Howling noise from the wheel
- Howling noise will increase when speeding up
- Two different howling noise from the same wheel

HOW FAULT FINDING WAS DONE

- Removed both rear wheels and inspected the wheel hubs. Wiggled the wheel discs back and forth. The left wheel disc attached to the wheel hub demonstrated a play while the right whee l disc did not Inspected around the bearing in the middle of
- the hub on both wheels. The left wheel hub was found to be leaking grease. Hence the cause of problem.

WHAT CAUSED THE FAULT

- Tear and wear of the wheel bearing

TH E FIX

- Replaced the rear left wheel hub with new one

OPTIONAL FIX

- Replace both the right and left wheel hubs with new ones

CHAPTER 5

GENERAL PROBLEMS

(91) KEY WARNING SYMBOL FLUSHING

- Veh icle Make: Mazda
- Model: Axela
- Year: 2003-2006
- Engine Code: LF
- Steering Wheel: Right Hand Drive
- Tra nsmission: Automatic

PROBLEM DESCRIPTION BY VEHICLE DRIVER

The instrument cluster was replaced as it had burnt components on it. The replacement cluster started flushing the key warning symbol continuously and car could only crank but not starting.

SYMPTOMS

- All lights on the instrument cluster come ON as normal but the key warning sign keeps on flushing
- Cranks but no START

HOW FAULT FINDING WAS DONE

The histo ry of the car is that the instrument cluster was replaced. The instrument cluster on this car is part of the immobilizer system and it therefore need to be programmed after replacement.

WHAT CAUSED THE FAULT

- Replacement instrument cluster came from another car

THE FIX

- Programmed the ECM and the immobilizer. Programming was stated 'SUCCESSFUL' but the key warning symbol did not stop flushing and hence car did not START. Then selected the option of 'Add Key' on the key programming equipment and this t ime the key warning symbol stopped flushing and the car STARTED.

(92) AIR CONDITION NOT WORKING

- Vehicle Make: VW
- Model: Polo (9N1/9N2/9N3)
- Year : 2005-2009
- Engine Code: BKY
- Steering Wheel: Right Hand Drive
- Transmission: Automatic

P ROBLEM DESCRIPTION BY VEHICLE DRIVER

There is no air conditioned air coming out from the A/C vents since the other mechanic worked on the car.

SYMPTOMS

- A/C control switches are able to switch ON
- No A/C refrigerant in the A/C pipes

HOW FAULT FINDING WAS DONE

- Teste d for pressure in the A/C pipes with pressure gauges and found no pressure. Hence the cause of the problem

WHAT CAUSED THE FAULT

- Loss of A/C refrigerant during previous mechanical works done on the car by other mechanic

THE FIX

- Refilled A/C refrigerant

(93) AIR CONDITIONER NOT WORKING

- Vehicle Make: Honda
- Model: Fit
- Year: 2002-2008
- Engine Code: L15A1
- Stee ring Wheel: Right Hand Drive
- Transmission: Automatic

PROB LEM DESCRIPTION BY VEHICLE DRIVER

The A/C refrigerant was refilled but there's no air conditioned air coming out from the A/C vents

SYMPTOMS

- A/C controls are able to switch ON
- A/C refrigerant has been recharged

HOW FAULT FINDING WAS DONE

- Tested for A/C pressure in the A/C pipes and found that pressure was ok
- Tested for power supply to A/C compressor after A/C system is switched ON but found no power. Hence the cause of the problem
- Traced the A/C compressor power supply cable and found it partially disconnected from its connector, hence the cause of the problem.

WHAT CAUSED THE FAULT

- Disc onnected power supply cable from the A/C compressor

THE FIX

- Reconnected the A/C power supply connector firmly to the A/C compressor connector

(94) BATTERY SMELLING AND BOILING

- Vehicle Make: VW
- Model: golf 6 TSI
- Year: 2008-2013
- Engine Code: CAXA
- Steering Wheel: Right Hand Drive
- Transmission: Automatic

PROBLEM DESCRIPTION BY VEHICLE DRIVER

Bad smell is coming out from the battery and the battery can be seen boiling

SYMPTOMS

- Battery very hot
- Boiling acid coming out of battery on its side
- Bad smell coming out from the battery

HOW FAULT FINDING WAS DONE

- Tested for battery overcharging (ok)
- Tested for battery under charging (ok)
- The Charging System proved to be ok, hence the battery is gone bad

WHAT CAUSED THE FAULT

- Battery gone bad due to age

SYMPTOMS

- Replaced the battery with a new

(95) START AND THEN IT STOPS IMMEDIATELY

- Vehicle Make: Mercedes-Benz
- Mo del: C-CLASS (203) C180K
- Year: 2002-2008
- Engine Code: 271. 946
- Steering Wheel: Right Hand Drive
- Tran smission: Automatic

PROBLEM DESCRIPTION BY VEHICLE DRIVER

When you crank the car, it starts very well but will immediately switch O FF

SYMPTOMS

- Car starts then it switches off immediately or just after one second

HOW FAULT FINDING WAS DONE

- Test ed fuel pressure (ok) Removed the downstream oxygen sensor from the exhaust pipe then started the car, car started without switching OFF. Hence the exhaust blocked on the catalytic converter after the downstream oxygen sensor.

WHAT CAUSED THE FAULT

- Bro ken catalytic converter particles causing blockage

THE FIX

- Unblocked the exhaust

OPTIONAL FIX

- Replace the catalytic converter with another one

(96) FRONT PASSENGER WINDOW NOT WORKING

- Vehicle Make: Mazda
- Model: Demio
- Year : 2007-2010
- Engine Code: ZJ
- Steering Wheel: Right Hand Drive
- Transmission: Automatic

PROBLEM DESCRIPTION BY VEHICLE DRIVER

The window in the front passenger door does not operate either by its switch or the switch on driver's door.

SYMPTOMS

- No sound of window motor running after switching ON its door switch or driver's door switch

- The window is fully closed, can't roll down by pressing any of its sw itches

HOW FAULT FINDING WAS DONE

- Teste d for 12 volts power to front passenger door switch when ignition switch is switched ON (OK)
- Tested for ground on front passenger window switch when it switches ON. (no ground),hence the cause of the problem.

WHAT CAUSED THE FAULT

- Ground missing at the switch on passenger door

THE FIX

- Restored ground to the front passenger window switch

(97) DRIVERS DOOR SWITCHES ARE NOT WORKING

- Vehicle Make: Mercedes Benz
- Model: S-CLASS (220) S320
- Year: 1998-2006
- Eng ine Code: 112. 944
- Steering Wheel: Right Hand Drive.
- Transmission: Automatic

PROBLEM DESCRIPTION BY VEHICLE DRIVER

Before the vehicle was parked, everything in the car was working just fine. Then car was parked for about three weeks without starting it. Suddenly problems emerged from no were as follows: the engine coolant radiator fan stopped working, all the four windows stopped working, driver seat adjustment stopped working, the boot lid stopped waking, all switches on driver's door stopped r esponding, the passenger's front window could only work from the switch on its door

SYMPTOMS

- Radiator fun not working
- All switches on driver's door not responding
- Only front passenger's window working by switch on its door
- Both front seats adjustment not working
- Side mirror not unfolding

HOW FAULT FINDING WAS DONE

- Battery negative disconnected Checked fuses on the power
- distribution box (Power Distribution is on foot floor on the right hand side).Fuse rated 150 amps on third position was found not to have continuity (blown fuse), hence cause of the problem

WHAT CAUSED THE FAULT

- Over current drawn through fuse rated 150 amp (This fuse is on the power distribution box on the foot floor of the driver side)

THE FIX

- Replaced the blown fuse with 150 amp fuse of same type

(98) ALL THE FOUR WINDOWS NOT WORKING

- Vehicle Make: Honda
- Mod el: Fit
- Year: 2001-2008
- Engine Code: L13
- Steering Wheel: Right Hand Drive
- Tran smission: Automatic

PROBLEM DESCRIPTION BY VEHICLE DRIVER

The car was taken for panel beating and all the windows were working. After panel beating was done I realized that all the windows were no longer working. All windows were working before pa nel beating works.

SYMPTOMS

- All four windows not working

HOW FAULT FINDING WAS DONE

- Checked the fuses for the windows (ok) Checked the constant
- power supply to the windows relay (ok) Check ed the earth to
- the windows relay (ok) Checked the ignition switch ON power
- supply to the windows relay. Ignition switch ON power supply to the window relay was found missing. Fuse box with window relay and fuse is under the steering wheel.

WHAT CAUSED THE FAULT

- Faulty fuse box

THE FIX

- Tapped ignition switch ON 12V positive power from the nearest and safest wire around the fuse box to supply window relay

OPTIONAL FIX

- Replace the fuse box under the steering wheel

(99) FRONT WIPERS NOT WORKING

- Vehicle Make: Audi
- Mod el: A3 (FSI)
- Year: 2005-2008
- Engine Code: BLR
- Steering Wheel: Right Hand Drive
- Transmission: Automatic

PROBLEM DESCRIPTION BY VEHICLE DRIVER

The wipers stopped working just by themselves. They were working at some point but then suddenly just stopped.

SYMPTOMS

- Wipers not working

HOW FAULT FINDING WAS DONE

- Disconnected the wiper motor from its power supply connector
- Tested for power supply on the connector (ok)
- Tested for earth on the connector (ok)
- Tested for voltage on the small wire on the connector (ok)
- All t he test proved to be OK hence the wiper motor assembly is faulty

WHAT CAUSED THE FAULT

- Tear and wear of wiper motor assembly

TH E FIX

- Replaced the wiper motor assembly

(100) ALL THE FOUR WINDOWS NOT WORKING

- Vehicle Make: Audi
- Model: A3 (FSI)
- Year: 2005-2008
- Engine Code: BLR
- Steering Wheel: Right Hand Drive
- Transmission: Automatic

PROB LEM DESCRIPTION BY VEHICLE DRIVER

All the windows were working just fine but suddenly they all stopped working

SYMPTOMS

- Driver's window not responding to switch
- Fro nt passenger's window not responding to any switch
- Rear right window not responding to any switch
- Rear left window not responding to any switch
- There is no winding sound from any window motors when switch is activated
- Cen tral locking not responding to remote control
- Central locking only responds to manual key

HOW FAULT FINDING WAS DONE

- Scan ned the gateway control module and retrieved trouble code: 01330: central control module for central convenience (J393) voltage supply too low Teste d for 12 power supply to drivers control module (OK) Tested for ignition switch ON power supply to drivers windows control module (OK) Tested for earth or ground to driver's window control module (OK) When DTC 01330 is cleared it immediately comes back hence the cause of the problem

WHAT CAUSED THE FAULT

- Driver's door window control module had stopped responding due to low battery voltage

THE FIX

- Disconnected the battery terminals then removed connectors from the driver's window control module for some minutes so that the module can reset itself and then reconnected the connectors back unto the module.

(101) ENGINE OIL WARNING LIGHT APPEARING ON DASHBOARD

- Vehicle Make: Audi
- Mo del: A3 (FSI)
- Year: 2005-2008
- Engine Code: BLR
- Steering Wheel: Right Hand Drive
- Tra nsmission: Automatic

PROBLEM DESCRIPTION BY VEHICLE DRIVER

The amber engine oil warning light appeared on the dashboard suddenly. Oil switch on the oil filter was replaced with a new one recently. Also the other oil switch on the engine oil sump was also replaced with a new one. Vehicle was also serviced but the amber engine oil light keeps showing on the dashboard.

SYMPTOMS

- Amber oil warning light stays ON permanently on the dashboard

HOW FAULT FINDING WAS DONE

- Diagnosed the car with diagnostic scanner and retrieved trouble code:
    - P0196: Engine oil temperature sensor-range/performance
- Unplugged connector from engine oil temperature sensor
- Tested the ground to the sensor (OK)
- Tested 12 volts power supply to the sensor (NO 12V), hence the cause of the problem

WHAT CAUSED THE FAULT

- 12 volts power supply to the engine oil temperature sensor missin g

THE FIX

- Restored the 12 volts power supply to the engine oil temperature sensor

(102) SPEEDO METER NOT WORKING

- Vehicle Make; VW
- Model: Polo (9N1/9N2/9N3)
- Year: 2005-2009

- Engine Code: BKY
- Stee ring Wheel: Right Hand Drive
- Transmission: Automatic

PROB LEM DESCRIPTION BY VEHICLE DRIVER

Suddenly speedometer stopped working

SYMPTOMS

- ABS warning light appearing on the dashboard
- Speedometer not working

HOW FAULT FINDING WAS DONE

- Tested for 12 volts constant power supply to the ABS control module (ok)
- Teste d for ignition ON 12 volts power supply to ABS control module. It was found missing. Hence causing ABS warning lamp on dashboard. If the ABS control module is not working/not switching ON the speedometer will not be responding on this car.

WHAT CAUSED THE FAULT

- No 12 volts constant power supply to ABS control module

THE FIX

- Restored ignition ON 12 volts power supply to ABS control module

(10 3) REMOTE CONTROL NOT WORKING

- Vehicle Make: Golf
- Mod el: GOLF V (GTI)
- Year: 2006-2008
- Engine Code: BWA
- Steering Wheel: Right Hand Drive
- Tran smission: Automatic

PROBLEM DESCRIPTION BY VEHICLE DRIVER

The remote control is not responding to lock or unlock of the central locking system. The manual key is able to lock and unlock the centr al locking system.

SYMPTOMS

- Red indicator light not appearing on the remote control when pressing any button

HOW FAULT FINDING WAS DONE

- Replaced the remote battery with a new battery but still red indicator light not appearing when pressing any button.
  Henc e the remote control is malfunctioned.

WHAT CAUSED THE FAULT

- Tear and wear of the remote control

T HE FIX

- Replaced the remote control with new one and programed it usin g diagnostic equipment

(104) FRONT LEFT WINDOW NOT WORKING

- Vehicle Make: Toyota
- Model: HILUX SURF
- Year : 1997-2003
- Engine Code: 3RZ-FE
- Steering Wheel: Right Hand Drive
- Transmission: Manual

PROBLEM DESCRIPTION BY VEHICLE DRIVER

The fron t left window does not respond either from the switch on the driver's door or the switch on its door.

SYMPTOMS

- Window not opening or closing
- No sound of window motor when switch is activated

HOW FAULT FINDING WAS DONE

- Tested for power supply to the window switch on front left door. 12 volts Power supply was missing hence the cause of the problem.

WHAT CAUSED THE FAULT

- Broken wire for 12 volts power supply somewhere in the wirin g harness

THE FIX

- Restored the 12 volts power supply to the window switch on thefront left door

(105) FUEL GUAGE ENDS AT HALF LEVEL

- Vehicle Make: NISSAN
- Model: X-TRAIL
- Engine Code: MR2
- Year : 2007-2011
- Steering Wheel: Right Hand Drive
- Transmission: Automatic

PROBLEM DESCRIPTION BY VEHICLE DRIVER

The fuel gauge does not go beyond half level even when the fuel is full in the tank. It only ends at half level.

SYMPTOMS

- Fuel gauge only ends at half level

HOW FAULT FINDING WAS DONE

- Disabled the power supply to fuel pump
- Removed the complete fuel pump housing from the tank.

- While fuel pump with housing is removed from the tank, conne cted it back on its power connector but the fuse for the fuel pump removed. Then turned ignition switch ON and

  starte d sliding the fuel gauge floater up and down to see if the gauge on the dashboard is responding. The dashboard gauge responded accordingly even up to full level, hence the problem was floater movement obstructed while inside the tank.

WHAT CAUSED THE FAULT

- Fuel pump not installed properly inside the tank by previous mechanic

THE FIX

- Reinstalled the complete fuel pump housing and positioned the fuel gauge sender unit floater nicely so it can move freely while in the tank

(106) DOORS NOT UNLOCKING OR LOCKING BY KEY OR REMOTE CONTROL

- Vehicle Make: VW
- Model: GOLF V
- Year: 2004
- Engin e Code: A XW
- Steering Wheel: Right Hand Drive
- Transmission: Automatic

PRO BLEM DESCRIPTION BY VEHICLE DRIVER

Doors cannot unlock or lock by manually using the key. The remote control is also not responding.

SYMPTOMS

- Manual key not unlocking or locking all the doors.
- Rem ote control not responding

HOW FAULT FINDING WAS DONE

- Rem oved the driver's door panel. Removed door key cylinder to inspect it. The key cylinder was
- en. Hence the cause of key not manually locking or brok
- unlocking the doors. The red indicator light on the remote control shows when pressing any button but no response to the central locking system hence remote control needs to be programed

WHAT CAUSED THE FAULT

- The key door cylinder broken due to being used very often since remote was not responding

- Remote control was not programmed after replacing the old one

TH E FIX

- Programmed the remote control so the doors can lock or unloc k using remote control

OPTIONAL FIX

- Also repair or replace the door lock cylinder

(107) NO CRANK

- Vehicle Make: Land Rover
- Model: Freelander
- Year : 2000-2006
- Engine Code: 18K
- Steering Wheel: Right Hand Drive
- Transmission: Manual

PROBLEM DESCRIPTION BY VEHICLE DRIVER

The remote control got lost so the vehicle is not cranking

SYMPTOMS

- Dashboard comes ON as normal
- No crank

H OW FAULT FINDING WAS DONE

- The v ehicle driver mentioned that the remote control had been lost, hence the cause of vehicle not cranking

WHAT CAUSED THE FAULT

- The remote control is part of the immobilizer system of this vehicle

THE FIX

- Reprogrammed new remote control

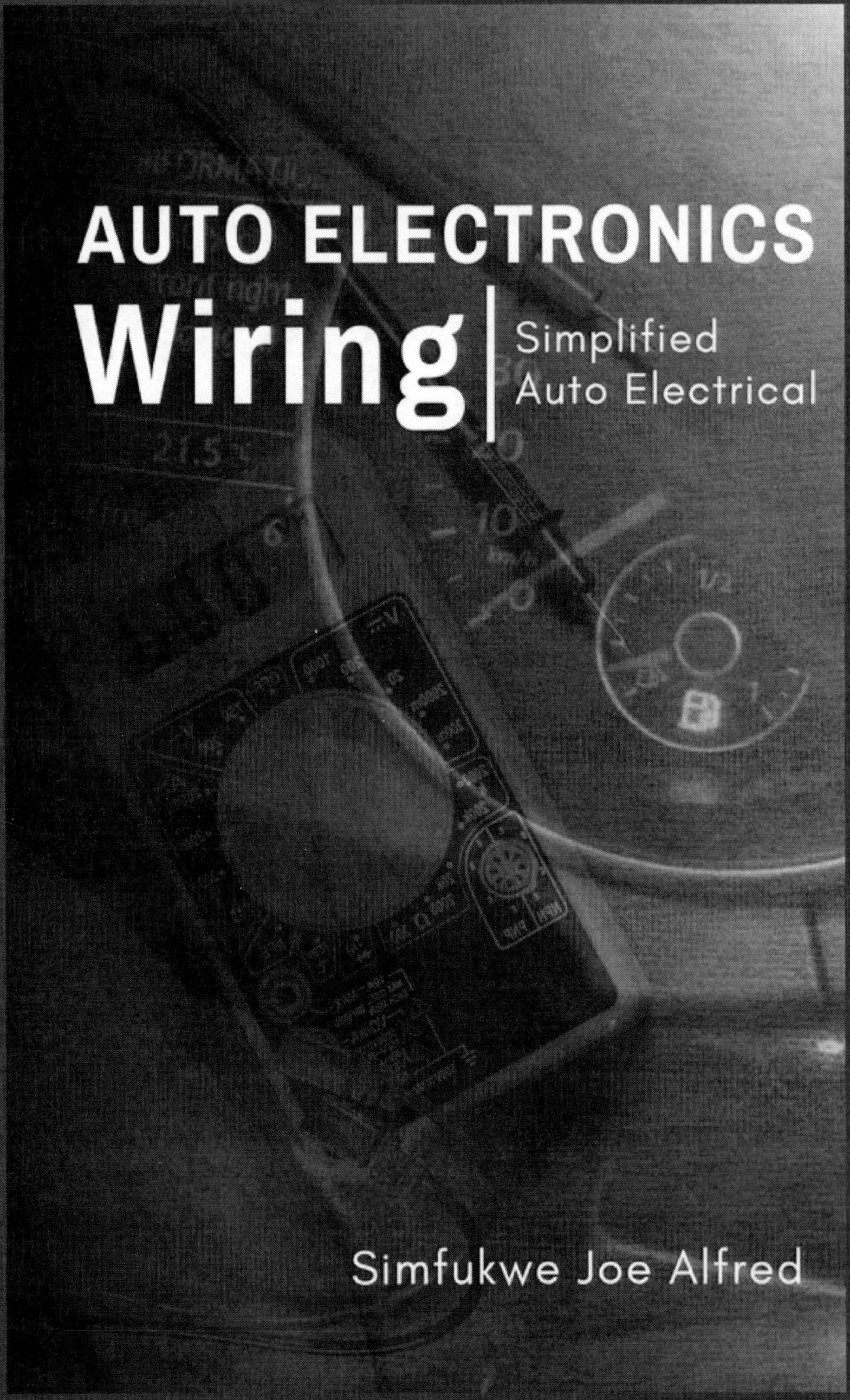
AUTO ELECTRONICS
Wiring
Simplified
Auto Electrical
Simfukwe Joe Alfred

Made in United States
Cleveland, OH
06 November 2026

47647242R00095